PROPERTIES OF ENGINEERING MATERIALS

For John

PROPERTIES OF
ENGINEERING MATERIALS

J. B. Moss M. Met., AIM, M. Weld I., Mappin Medallist

Head of the Department of Pure and Applied Science,
Consett Technical College

A DIVISION OF
THE **CHEMICAL RUBBER** CO.
CLEVELAND, OHIO

INTERNATIONAL SCIENTIFIC SERIES

English edition published in 1971 by
Butterworth & Co (Publishers) Ltd.

©, Butterworth & Co. (Publishers) Ltd., London 1971

Published in the U.S.A. by
The Chemical Rubber Co.,
18901 Cranwood Parkway,
Cleveland, Ohio 44128

Library of Congress Catalog Card Number 72 – 173 785

PREFACE

Most present-day courses in Engineering include a study of materials under headings such as 'Materials Science' or 'Properties of Materials'. This is perhaps a reflection of the feeling that the majority of future advances in engineering technology will come through developments in the materials of engineering rather than in basic engineering science, a fact which has been amply demonstrated in the development of the gas turbine.

This present book does not set out to be a course in Materials Science, hence its chosen title. Instead, it aims to discuss the properties and behaviour of materials, their response to environment, and to relate these properties to unifying principles.

It is primarily intended for Mechanical and Production Engineers at advanced level who are working on HND, CEI, and CNAA degree courses. It should also be useful in the new HNC scheme in Engineering. The book is not intended to be a complete statement of material properties in itself, rather should it be regarded as an adjunct to lectures given on this topic.

SI units have been used throughout the text but conversion tables are included to ease the transition into this system.

CONTENTS

Chapter 1

ATOMS AND MOLECULES

The first three chapters of this book provide the basic principles of structure and behaviour which are needed for an appreciation of later work. In this chapter a review is given of those properties of atoms and molecules which are used in later chapters to explain engineering properties. An attempt is made to put the one hundred or so known chemical elements into some sort of order and perspective and to point out the fundamental differences between them.

DESCRIPTION OF THE ATOM

The bulk properties of an engineering material are simply the properties of the community of atoms of which it is composed. The relevant features concerning atomic structure and properties are therefore outlined here.

The classical description of the atom visualises it as a tiny solar system consisting of:

(a) A central nucleus carrying protons and neutrons and having dimensions of the order of 1×10^{-10} m. The nucleus is positively charged since each proton, of mass $1 \cdot 6724 \times 10^{-27}$ kg, carries a positive charge of $1 \cdot 602 \times 10^{-19}$ C. The neutron is a neutral particle of mass $1 \cdot 6747 \times 10^{-27}$ kg.

(b) Negatively charged particles in orbit around the nucleus. These electrons have a mass of $9 \cdot 1066 \times 10^{-31}$ kg and carry a negative charge of $1 \cdot 602 \times 10^{-19}$ C. The number of electrons in orbit is the same as the number of protons on the nucleus in the normal atom.

In this concept of the atom, stability was supposed to be the result of balance between the centrifugal force on the orbiting electrons and the electrostatic attraction between nucleus and electrons. If this concept is analysed, however, it becomes apparent that the electrons must be spiralling down towards the nucleus and that the atom should be radiating energy. In fact, neither process occurs in the normal atom and so there must be something inherently incorrect in trying to visualise an atom as a miniature solar system.

To overcome the difficulties with this classical model of the atom, Bohr introduced, in 1913, the idea of quantitised electron energy states which he based on work by Planck.

Planck had already shown that when an atom was energised by outside energy, it did not take up this energy smoothly and continuously, but rather absorbed it in a number of separate packets or quanta. This means that an atom can only exist in certain energy states and the permitted energy levels were given as $\varepsilon = h\nu$, where h is Planck's constant, and ν is the frequency of atomic vibration.

Atoms apparently could take up packets or quanta of energy equal to whole number multiples of ε but states such as $\frac{1}{2}\varepsilon$, $\frac{1}{4}\varepsilon$ were forbidden.

Bohr applied these ideas to the energies of electrons within atoms and postulated that electrons in atoms could rotate only in certain permitted orbits, and that when in these orbits the electron could orbit without radiating energy. These permitted energy levels or orbits were termed stationary states, and in such a state the electron must have a whole number of energy quanta and will be described by a relationship $mvr = (nh)/(2\pi)$, where m, v are the mass and velocity of the electron, r is the radius of electron orbit, and n is the quantum number which is always a whole number.

The total energy of an electron in a stationary state was given as

$$E = -\frac{K}{n^2} \quad \text{eV}$$

The value of K for the normal hydrogen atom is 13·58 so it should be possible to predict the permitted stationary states. Thus, when $n = 1$ the hydrogen atom is in its normal or ground condition.

By adding electrical or thermal energy, it is quite possible to excite the atom so that the electron is lifted into energy states for which n is greater than 1·0. However, n must always be a whole number.

Thus, when $n = 1$ $\quad E = -13\cdot58$ eV (1 eV $= 1\cdot602\times10^{-19}$ J).

$n = 2$ $\quad E = -\ 3\cdot39$ eV

$n = 3$ $\quad E = -\ 1\cdot57$ eV

$n = 4$ $\quad E = -\ 0\cdot85$ eV

An increase in n corresponds to an increase in electron energy and this involves an increase in radius r.

Thus, an energised electron will move out of its normal orbit and will jump the gap into the next highest permitted zone. The idea is also conveyed that if an atom contains a number of electrons of varying energies, then these electrons will be disposed around the nucleus at various levels, the levels crowding together as n increases.

In the above description of the atom, it is assumed that the electron is a particle of definite mass and so it should be possible to define its speed and position using ordinary Newtonian mechanics.

This, in fact, is impossible to do, since any experiment which sets out to measure electron speed or position would alter the properties of the electron. So, while Newtonian mechanics may be serviceable in describing the properties of large masses, they cannot be used to analyse particles of atomic dimensions.

A more general description of the atom makes use of a form of mechanics known as wave mechanics, which considers only the energies associated with moving masses. Such a system gives up any attempt to regard the electron as a finite particle but instead considers it as a packet of wave energy vibrating on a definite wavelength. An electron can thus be visualised as an energy 'cloud' blanketing the nucleus and this cloud can be thought of as being most dense where the probability of finding the electron is greatest.

The equations produced by wave mechanics are found to give sensible solutions only with certain definite values of electron energy. Energy levels between these values are found to be impossible.

The wave mechanical approach, therefore, leads to similar conclusions to the Bohr approach but is more generally applicable and avoids the errors inherent in treating the electron as a particle. Fig. 1.1 represents a hydrogen atom in its ground state. A probability function P, as calculated using wave mechanics, is plotted against the distance r from the nucleus. The density of the electron cloud at any point is proportional to the probability of finding the electron at that point. There is maximum probability of finding the electron 0.53×10^{-10} m from the nucleus, but since probability is a continuous function we must be prepared to find the electron other than at $r = 0.53 \times 10^{-10}$ m. The electron energy states nearest to the nucleus are usually spherical in shape but the probability clouds for outer levels may be anything but spherical. Transfer of electron energy from one probability state or level to another simply involves an instantaneous change in cloud den-

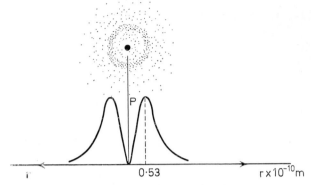

Fig. 1.1. The electron in the normal hydrogen atom

sity. The transfer from one state to another must still occur without traversing intermediate levels. The effect is similar to the way in which a wind instrument can change from one pitch (wavelength) to another without traversing the pitches in between.

ATOMIC WEIGHT AND ATOMIC NUMBER

The real weight of an atom is mainly due to the protons and neutrons on the nucleus, since the weight of the electron is only about 0·0005 of that of the nucleus.

The atomic weight of an element is not this real weight, but simply a comparative value given by assuming that the weight of an oxygen atom can be represented by the number 16. On such a scale, the number which must be assigned to hydrogen is 1·008. Such atomic weights can be expressed in any units.

The real weight of an atom is related to the atomic weight and can be found using Avogadro's number, i.e. weight of one atom equals

$$\frac{\text{atom weight in grammes}}{6\cdot02\times10^{23}}$$

The Avogadro number is thus the number of atoms per gramme atomic weight.

The *atomic number* Z is simply the number of protons on the nucleus of the atom. In the normal atom, Z is the same as the number of electrons in orbit.

The *mass number* A is the sum of protons and neutrons on a nucleus and as such is not quite the same as the weight of the atom.

In many cases atoms will have identical atomic numbers Z but different mass numbers A. Such atoms will thus have variable neutron content $(A - Z)$ and are referred to as *isotopes*.

In fact, bulk samples of all materials are composed of collections of isotopes and so what is called the atomic weight is really the average mass number of the common isotopes of which the material is composed.

ELECTRON CONFIGURATION

The differences between different chemical elements arise simply because of differences in the number of protons, neutrons and electrons which the atoms contain.

It would therefore be useful to know about the dispositions and numbers of fundamental particles and, since many of the important properties of elements are governed by the way in which the electrons are arranged in the atoms, a study of electron configuration is indicated.

Whether the electron is regarded as a solid particle or as wave energy, it will have two simultaneous motions: (*a*) a spin about its own axis, and (*b*) a revolution around the nucleus. And, since it constitutes a moving charge, it will be associated with a tiny magnetic moment. There are limitations on the way in which electrons are disposed around a nucleus and these limitations are embodied in the Pauli Exclusion Principle which states that no electron energy level may contain more than two electrons, and for these two to be accommodated simultaneously they must have opposite spins.

The energy state of an electron is specified by a number of symbols:

n, the total quantum number indicating the total energy of the electron. Its value increases as orbit diameter increases.

l, a secondary value which measures the angular momentum of an electron relative to the nucleus (or its wavelength if one regards the electron as wave energy). The value of l may be between 0 and $(n-1)$ but is never greater than 3. A condition $l = 0$ means that the electron is in such a condition that it could be moving either way around the nucleus.

This range $l = 0$ to $l = (n-1)$ imposes limitations on the value of l. Thus an electron at an energy level for which $n = 2$ could only have l values of 0 or 1.

By convention, the states $l = 0, 1, 2, 3$ are called s, p, d, f, and so an expression $3d^1$ indicates that 1 electron is present at an energy level $n = 3$ and has a momentum for which $l = 2$.

The number of energy states in an atom is limited practically by the fact that, at some stage, electron configuration becomes so complex that the atom becomes unstable and disintegrates to less complex forms. Hence, there is a limit to the number of chemical elements available to us. Every atom, no matter of which chemical element, contains all possible electron energy states, but in the lighter elements only the first few available levels are actually occupied by electrons. Higher energy states, i.e. levels further away from the nucleus, become occupied as atomic weight increases or if an atom is excited by external electrical or thermal energy.

The possible available electron energy levels in an atom are indicated in Table 1.1. These are the theoretical energy levels arranged in an energetically logical order, but in real atoms they may not appear in this order.

Table 1.1. POSSIBLE ELECTRON ENERGY LEVELS IN ATOMS

'Shell' designation	*Energy conditions*			*Total capacity for electrons*
Shell Q	$n = 6$	$l = 2$	$6d$	10
	$n = 7$	$l = 1$	$7p$	6
	$n = 7$	$l = 0$	$7s$	2
Shell P	$n = 4$	$l = 3$	$4f$	14
	$n = 5$	$l = 2$	$5d$	10
	$n = 6$	$l = 1$	$6p$	6
	$n = 6$	$l = 0$	$6s$	2
Shell O	$n = 4$	$l = 2$	$4d$	10
	$n = 5$	$l = 1$	$5p$	6
	$n = 5$	$l = 0$	$5s$	2
Shell N	$n = 3$	$l = 2$	$3d$	10
	$n = 4$	$l = 1$	$4p$	6
	$n = 4$	$l = 0$	$4s$	2
Shell M	$n = 3$	$l = 1$	$3p$	6
	$n = 3$	$l = 0$	$3s$	2
Shell L	$n = 2$	$l = 1$	$2p$	6
	$n = 2$	$l = 0$	$2s$	2
Shell K	$n = 1$	$l = 0$	$1s$	2

NUCLEUS

By arranging the chemical elements in order of increasing atomic number Z, the pattern of the build-up of the atoms is clearly seen. Table 1.2 shows clearly how one element differs from another. Some general conclusions can be drawn from this table:

1. Each energy level eventually reaches a saturation value and thereafter remains constant.

2. With simple atoms at the beginning of the table, a new energy level is not occupied until the previous levels are full.

3. The chemically inert elements such as neon, argon and xenon have outermost energy levels which are either saturated, as with helium, or contain 8 electrons. In fact, all the inert gases have saturated outer shells, since helium has a final s state which is saturated with 2 electrons, while the rest have final p states which are saturated with 6 electrons. We might therefore associate an outer shell containing 8 electrons with particular stability.

4. Metallic elements are characterised by having only a few electrons in their outer shells, usually less than 4. For example:

calcium	Ca	2—8—8—2
iron	Fe	2—8—14—2
copper	Cu	2—8—18—1

The outermost shell of electrons in metals is called the valency shell and the number of valency electrons which a metal has governs its chemical reactivity.

5. Non-metallic elements are characterised by comparatively large numbers of electrons in the outer shells. For example:

chlorine	Cl	2—8—7
sulphur	S	2—8—6
iodine	I	2—8—18—18—7

In such cases the valency number is not coincident with the number of electrons in the outer shell.

6. Elements having 4 or 5 electrons in their outer shells could obviously have both metallic and non-metallic properties. These are elements such as carbon, silicon, germanium and bismuth, some of which are important as semiconductor devices.

Table 1.2. ELECTRON CONFIGURATION

			K	L		M			N				O				P			Q
$n =$			1	2		3			4				5				6			7
$l =$			0	0	1	0	1	2	0	1	2	3	0	1	2	3	0	1	2	0
			$1s$	$2s$	$2p$	$3s$	$3p$	$3d$	$4s$	$4p$	$4d$	$4f$	$5s$	$5p$	$5d$	$5f$	$6s$	$6p$	$6d$	$7s$
Z	Element																			
1	Hydrogen	H	1																	
2	Helium	He	2																	
3	Lithium	Li	2	1																
4	Beryllium	Be	2	2																
5	Boron	B	2	2	1															
6	Carbon	C	2	2	2															
7	Nitrogen	N	2	2	3															
8	Oxygen	O	2	2	4															
9	Fluorine	F	2	2	5															
10	Neon	Ne	2	2	6															
11	Sodium	Na	2	2	6	1														
12	Magnesium	Mg	2	2	6	2														
13	Aluminium	Al	2	2	6	2	1													
14	Silicon	Si	2	2	6	2	2													
15	Phosphorus	P	2	2	6	2	3													
16	Sulphur	S	2	2	6	2	4													
17	Chlorine	Cl	2	2	6	2	5													
18	Argon	A	2	2	6	2	6													

19	Potassium	K	2	2	6	2	6		1			
20	Calcium	Ca	2	2	6	2	6		2			
21	Scandium	Sc	2	2	6	2	6	1	2			
22	Titanium	Ti	2	2	6	2	6	2	2			
23	Vanadium	V	2	2	6	2	6	3	2			
24	Chromium	Cr	2	2	6	2	6	5	1			
25	Manganese	Mn	2	2	6	2	6	5	2			
26	Iron	Fe	2	2	6	2	6	6	2			
27	Cobalt	Co	2	2	6	2	6	7	2			
28	Nickel	Ni	2	2	6	2	6	8	2			
29	Copper	Cu	2	2	6	2	6	10	1			
30	Zinc	Zn	2	2	6	2	6	10	2			
31	Gallium	Ga	2	2	6	2	6	10	2	1		
32	Germanium	Ge	2	2	6	2	6	10	2	2		
33	Arsenic	As	2	2	6	2	6	10	2	3		
34	Selenium	Se	2	2	6	2	6	10	2	4		
35	Bromine	Br	2	2	6	2	6	10	2	5		
36	Krypton	Kr	2	2	6	2	6	10	2	6		
37	Rubidium	Rb	2	2	6	2	6	10	2	6		1
38	Strontium	Sr	2	2	6	2	6	10	2	6		2
39	Yttrium	Y	2	2	6	2	6	10	2	6	1	2
40	Zirconium	Zr	2	2	6	2	6	10	2	6	2	2
41	Niobium	Nb	2	2	6	2	6	10	2	6	4	1
42	Molybdenum	Mo	2	2	6	2	6	10	2	6	5	1
43	Technetium	Tc	2	2	6	2	6	10	2	6	6	1

continued overleaf

Table 1.2 continued

Z	Element		K (n=1)	L (n=2)		M (n=3)			N (n=4)				O (n=5)				P (n=6)			Q (n=7)
			1s	2s	2p	3s	3p	3d	4s	4p	4d	4f	5s	5p	5d	5f	6s	6p	6d	7s
44	Ruthenium	Ru	2	2	6	2	6	10	2	6	7		1							
45	Rhodium	Rh	2	2	6	2	6	10	2	6	8		1							
46	Palladium	Pd	2	2	6	2	6	10	2	6	10									
47	Silver	Ag	2	2	6	2	6	10	2	6	10		1							
48	Cadmium	Cd	2	2	6	2	6	10	2	6	10		2							
49	Indium	In	2	2	6	2	6	10	2	6	10		2	1						
50	Tin	Sn	2	2	6	2	6	10	2	6	10		2	2						
51	Antimony	Sb	2	2	6	2	6	10	2	6	10		2	3						
52	Tellurium	Te	2	2	6	2	6	10	2	6	10		2	4						
53	Iodine	I	2	2	6	2	6	10	2	6	10		2	5						
54	Xenon	Xe	2	2	6	2	6	10	2	6	10		2	6						
55	Cesium	Cs	2	2	6	2	6	10	2	6	10		2	6			1			
56	Barium	Ba	2	2	6	2	6	10	2	6	10		2	6			2			
57	Lanthanum	La	2	2	6	2	6	10	2	6	10		2	6	1		2			
58	Cerium	Ce	2	2	6	2	6	10	2	6	10	2	2	6			2			
59	Praseodymium	Pr	2	2	6	2	6	10	2	6	10	3	2	6			2			
60	Neodymium	Nd	2	2	6	2	6	10	2	6	10	4	2	6			2			
61	Promethium	Pm	2	2	6	2	6	10	2	6	10	5	2	6			2			

No.	Name	Symbol	1s	2s	2p	3s	3p	3d	4s	4p	4d	4f	5s	5p	5d	6s	6p	7s
62	Samarium	Sm	2	2	6	2	6	10	2	6	10	6	2	6		2		
63	Europium	Eu	2	2	6	2	6	10	2	6	10	7	2	6		2		
64	Gadolinium	Gd	2	2	6	2	6	10	2	6	10	7	2	6	1	2		
65	Terbium	Tb	2	2	6	2	6	10	2	6	10	9	2	6		2		
66	Dysprosium	Dy	2	2	6	2	6	10	2	6	10	10	2	6		2		
67	Holmium	Ho	2	2	6	2	6	10	2	6	10	11	2	6		2		
68	Erbium	Er	2	2	6	2	6	10	2	6	10	12	2	6		2		
69	Thulium	Tm	2	2	6	2	6	10	2	6	10	13	2	6		2		
70	Ytterbium	Yb	2	2	6	2	6	10	2	6	10	14	2	6		2		
71	Lutecium	Lu	2	2	6	2	6	10	2	6	10	14	2	6	1	2		
72	Hafnium	Hf	2	2	6	2	6	10	2	6	10	14	2	6	2	2		
73	Tantalum	Ta	2	2	6	2	6	10	2	6	10	14	2	6	3	2		
74	Tungsten	W	2	2	6	2	6	10	2	6	10	14	2	6	4	2		
75	Rhenium	Re	2	2	6	2	6	10	2	6	10	14	2	6	5	2		
76	Osmium	Os	2	2	6	2	6	10	2	6	10	14	2	6	6	2		
77	Iridium	Ir	2	2	6	2	6	10	2	6	10	14	2	6	7	2		
78	Platinum	Pt	2	2	6	2	6	10	2	6	10	14	2	6	8	2		
79	Gold	Au	2	2	6	2	6	10	2	6	10	14	2	6	10	1		
80	Mercury	Hg	2	2	6	2	6	10	2	6	10	14	2	6	10	2		
81	Thallium	Tl	2	2	6	2	6	10	2	6	10	14	2	6	10	2	1	
82	Lead	Pb	2	2	6	2	6	10	2	6	10	14	2	6	10	2	2	
83	Bismuth	Bi	2	2	6	2	6	10	2	6	10	14	2	6	10	2	3	
84	Polonium	Po	2	2	6	2	6	10	2	6	10	14	2	6	10	2	4	
85	Astatine	At	2	2	6	2	6	10	2	6	10	14	2	6	10	2	5	
86	Radon	Rn	2	2	6	2	6	10	2	6	10	14	2	6	10	2	6	
87	Francium	Fr	2	2	6	2	6	10	2	6	10	14	2	6	10	2	6	1

continued overleaf

Table 1.2 continued

		K	L		M			N				O				P			Q
n =		1	2		3			4				5				6			7
l =		0	0	1	0	1	2	0	1	2	3	0	1	2	3	0	1	2	0
		$1s$	$2s$	$2p$	$3s$	$3p$	$3d$	$4s$	$4p$	$4d$	$4f$	$5s$	$5p$	$5d$	$5f$	$6s$	$6p$	$6d$	$7s$
Z	*Element*																		
88	Radium Ra	2	2	6	2	6	10	2	6	10	14	2	6	10		2	6		2
89	Actinium Ac	2	2	6	2	6	10	2	6	10	14	2	6	10		2	6	1	2
90	Thorium Th	2	2	6	2	6	10	2	6	10	14	2	6	10		2	6	2	2
91	Protoactinium Pa	2	2	6	2	6	10	2	6	10	14	2	6	10	1	2	6	2	2
92	Uranium U	2	2	6	2	6	10	2	6	10	14	2	6	10	3	2	6	1	2
93	Neptunium Np	2	2	6	2	6	10	2	6	10	14	2	6	10	4	2	6	1	2
94	Plutonium Pu	2	2	6	2	6	10	2	6	10	14	2	6	10	6	2	6		2
95	Americium Am	2	2	6	2	6	10	2	6	10	14	2	6	10	7	2	6		2
96	Curium Cm	2	2	6	2	6	10	2	6	10	14	2	6	10	7	2	6	1	2
97	Berkelium Bk	2	2	6	2	6	10	2	6	10	14	2	6	10	8	2	6	1	2
98	Californium Cf	2	2	6	2	6	10	2	6	10	14	2	6	10	10	2	6		2

The electron configuration of the first few elements in Table 1.2 can be pictured using the concept of probability clouds:

Hydrogen H. $Z = 1$; 1 proton; 1 electron: a spherical cloud around the nucleus and having maximum density at a radius of 0.53×10^{-10} m from the nucleus.

Lithium Li. $Z = 3$; 3 protons+4 neutrons; 3 electrons: two electrons occupy the $1s$ state, while the third electron occupies the $2s$ state. Both clouds are spherical and have maximum densities at 0.2×10^{-10} m and 1.5×10^{-10} m.

Boron B. $Z = 5$; 5 protons+6 neutrons; 5 electrons: two electrons are in the spherical $1s$ cloud with maxi-

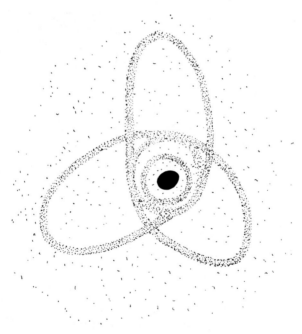

Fig. 1.2. Electron cloud patterns in the boron atom

mum probability at about 0.12×10^{-10} m. The three electrons in the L shell are contained in three lobe-shaped clouds, making angles of 120° with one another. Fig. 1.2 illustrates this configuration in boron.

Carbon C. $Z = 6$; 6 protons+6 neutrons; 6 electrons: an inner spherical $1s$ cloud contains 2 electrons. The outer levels consist of four lobes, each containing one electron, the lobes having their axes pointing towards the corners of a regular tetrahedron. This tetrahedral distribution of the outer electrons around the carbon nucleus means that a long chain of carbon atoms all chemically joined together will have the appearance of a three-dimensional zig-zag and will not be planar. Such chains occur in plastic materials and have a large effect on the properties of these materials.

The tetrahedral lobe configuration first established with carbon becomes saturated when each lobe contains 2 electrons. This occurs at neon, $Z = 10$. Beyond neon, further energy clouds are established and it is evident that patterns become extremely complicated.

THE PERIODIC CLASSIFICATION OF THE ELEMENTS

Table 1.2 divided the list of elements into seven sections or periods, each period coming to an end with an inert gas.

By rearranging the table it becomes possible to predict some of the properties of an element simply by considering its position in relation to its neighbours. One way of rearrangement is shown in Table 1.3 and is one example of a periodic classification of the elements.

The horizontal rows are called *periods* and the vertical columns are the *groups*.

In the periods, the elements are arranged in order of increasing atomic number Z and, by so doing, it is evident that within the groups the elements are now arranged in order of chemical similarity. Thus, silver Ag in group IB has good corrosion resistance and so it would be expected that copper and gold would also have this property.

In the A groups, elements are arranged according to the number of outer shell electrons. Hence carbon, silicon and germanium in group IVA all have 4 outer shell electrons and would, as a result, be expected to have similar chemical properties. The 0 group may be regarded as a special form of A group in which the outer electron level is saturated. This group includes the chemically inert elements, and the outer level containing 8 electrons could be regarded as

Table 1.3. PERIODIC CLASSIFICATION OF THE ELEMENTS

IA	IIA
H	
Li	Be
Na	Mg
K	Ca
Rb	Sr
Cs	Ba
Fr	Ra

IIIB	IVB	VB	VIB	VIIB	VIII			IB	IIB
Sc	Ti	V	Cr	Mn	Fe	Co	Ni	Cu	Zn
Y	Zr	Nb	Mo	Tc	Ru	Rh	Pd	Ag	Cd
Lanthanides	Hf	Ta	W	Re	Os	Ir	Pt	Au	Hg
Actinides									

Lanthanides La to Lu

Actinides Ac to Cf

IIIA	IVA	VA	VIA	VIIA	0
				H	He
B	C	N	O	F	Ne
Al	Si	P	S	Cl	Ar
Ga	Ge	As	Se	Br	Kr
In	Sn	Sb	Te	I	Xe
Tl	Pb	Bi	Po	At	Rn

being particularly stable. Chemical reaction between atoms, to produce molecules, may occur by exchange of outer level electrons between the atoms, and it is logical to assume that if such an exchange does occur it will occur in such a way as to leave the resultant parts of the molecule with the stable outer octet arrangement.

Thus sodium Na (2—8—1) may combine chemically with chlorine Cl (2—8—7) by giving its outer shell electron to the chlorine atom. Both sides of the resultant molecule will now have configurations ending in 8 and will be held together by electrostatic attraction.

Group IA elements can readily lose the single electron from their outer shells and VIIA elements can readily accept a single electron. Such elements would therefore be expected to be highly active chemically.

Chemical reactivity decreases with IIA and IIIA elements, since these must lose more outer shell electrons.

There will therefore come a stage at which chemical combination by loss of electrons becomes unlikely and this obviously occurs with IVA elements. These elements tend to undergo chemical combinations not by electron loss or acceptance but by electron sharing in order to reach the octet arrangement. This produces the co-valent bond which is discussed later and which forms the basis of the structures of engineering plastics and rubbers.

The B groups and group VIII are termed 'transition elements', since the build-up of electron configurations is not regular as with A group elements. In these B group elements, electrons can be taken up by levels below the outermost level, as is shown by Table 1.2. All these transition elements are metals and have a small number of electrons in the outermost level. Chemical reactivity in these elements cannot be related to the number of outer level electrons, as was the case with A group elements, since with the metals, loss of outer shell electrons does not leave a stable octet arrangement behind. Thus both copper and sodium have a single electron in the outer level but copper would not be predicted as having high chemical activity.

Most of the important chemical properties of elements are related to their electron configurations and so the usefulness of the periodic classification lies in its ability to predict likely properties of elements from a knowledge of the properties of its neighbours in the group.

Some physical properties such as freezing point and boiling point also follow a periodic relationship across the rows, i.e. within the periods.

BONDING BETWEEN ATOMS

Communities of atoms make up the bulk materials with which we are familiar. The atoms in such communities must be very strongly held together and it is the purpose of this section to outline the types of bonding forces which hold atoms together.

Any atom consists of a positively charged nucleus surrounded by a negatively charged electron cloud. Hence the relative positions which two atoms will take up must be a balance between the repulsion of their electron clouds and the attraction arising from their desire to change their outer level electron states to the stable octet arrangement.

Atoms in a solid body community will therefore take up definite positions in relation to each other and any attempt to alter this spacing will be strongly resisted. The value of the energy needed to move an atom completely away from its equilibrium position is a measure of the binding force between atoms. This force varies, depending on whether the bond is of the primary or secondary type.

Primary bonds

These are extremely strong and energies of the order of $4 \cdot 2 \times 10^8$ J/kg mol would be needed to cause disruption. The various types of primary bond are as follows:

Ionic bonds

This type of bond between atoms arises from the mutual attraction of positive and negative charges created by the transfer or acceptance of outer level electrons.

A molecule of sodium chloride NaCl could be visualised as shown in Fig. 1.3, i.e. the single valency electron from the sodium is given to the chlorine atom giving both parts of the molecule the stable outer octet arrangement. The transfer of the electron leaves the sodium with an excess positive charge, Na⁺, and the

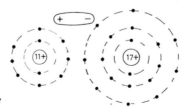

Fig. 1.3. The NaCl *molecule*

chlorine with an extra negative charge, Cl^-. These charged particles are known as *ions* and the molecule is held together by attraction between the charges on the ions.

Bonding becomes stronger as more electrons are transferred and this of course depends on the valency of the metal atom which is donating the electrons.

Such ionically bonded solids are capable of carrying an electric current if the ions can be made mobile by either melting or dissolving in water. Solutions of ionic solids are, therefore, electrolytes.

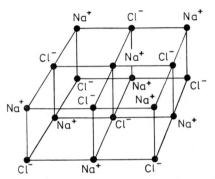

Fig. 1.4. Cubic arrangement of ions in sodium chloride

In bulk, the ions in an ionically bonded material will always tend to surround themselves with ions of unlike charge. This produces a regular three-dimensional array of ions. A material in which atomic or ionic arrangement is regular is classified as being crystalline. Fig. 1.4 illustrates the cubic arrangement of ions in sodium chloride. Many ceramic materials used in furnace refractories contain the ionic bond.

The co-valent or homopolar bond

In this type of bond, the atoms are held together since they share their outer level electrons in an attempt to reach stability.

The simplest co-valently bonded molecule is that of hydrogen. In this case, the two outer level electrons (one from each atom) are shared so that both atoms are associated with two electrons. This gives the stable helium configuration, and since a single pair of electrons is shared we could visualise the hydrogen molecule as having a single co-valent bond, i.e. H—H.

The oxygen molecule is also co-valently bonded. Each oxygen atom has 6 electrons in its outer level, disposed in four tetrahedrally arranged lobes. Two lobes contain 2 electrons each while the other two lobes contain only 1 electron each. On molecule formation, the single electron lobes from each atom are shared so that, for part of the time at least, each side of the molecule has 4 lobes each containing 2 electrons. This gives the stable octet arrangement and, since 2 pairs of electrons are involved in the sharing, the oxygen molecule has a double co-valent bond, i.e. $O\!\!=\!\!O$.

With hydrogen and oxygen, the nuclei involved are similar and electron sharing is equal. In most co-valently bonded compounds, however, this is not the case. For example, with carbon dioxide CO_2 the atoms involved are dissimilar and sharing is not equal. Carbon has 4 outer electron lobes each containing a single electron. An oxygen atom has two of its four lobes containing a single electron per lobe. The 4 electrons from a single carbon atom are shared with the single-electron lobes from each of two oxygen atoms, giving two double bonds, i.e. $O\!\!=\!\!C\!\!=\!\!O$. The 4 unsaturated lobes in the carbon atom permit a vast range of co-valent molecules to be built up by combination of carbon with other elements. These are the organic molecules and include living substances, fuels, plastics, etc.

The co-valent bond is extremely strong, as evidenced by the hardness and strength of diamond which consists of carbon atoms co-valently held together. The melting point of an element reflects the strength of the bond between the atoms of that element, since melting involves disruption of bonds by thermal energy.

Since the co-valent bond involves the sharing of electrons and since the shared electrons are associated equally with both sides of the molecule, then excess charges are not present. A co-valently bonded material would not, therefore, be expected to be either a good electrical or a good thermal conductor. This is true of many ceramic materials.

Silicon Si and germanium Ge, like carbon, appear in group IVA of the periodic classification and would thus also be expected to participate in co-valent bonding. These elements are also of interest since they are used in the manufacture of semiconducting devices such as transistors.

The metallic bond

Single atoms of metals, by definition, contain only a few electrons in their outer shells. Thus, in any aggregate of metal atoms, there is no possibility of stable octet arrangements being achieved. Nevertheless, atoms in bulk metal must be held together ver strongly, as is seen by the high fracture strength of metals.

The bonding present is peculiar to metals and involves all the valency electrons from all the atoms in the aggregate being shared equally between all atoms.

The common valency electron cloud so produced permeates the whole volume of the metal and is free to drift anywhere within the volume. A piece of metal must, therefore, be visualised as a collection of positively charged ions bonded together by a freely moving common cloud of negative valency electrons. The presence of free electrons explains the good thermal and electrical conductivity of metals.

Such a collection of positive ions embedded in a negative electron cloud will, because of the forces involved, be arranged in a regular array. The positive ions will take up definite positions in space relative to each other and hence metallic materials are crystalline in nature, as indicated in Fig. 1.5.

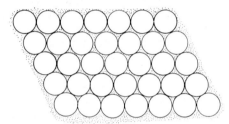

Fig. 1.5. Regular array of positively charged metal ions permeated by the valency electron 'cloud'

This type of bond occurs whenever the atoms would need to receive more than 4 electrons to give an inert gas configuration. It is the bond, therefore, which is present in all bulk materials to the left of group IVA in the periodic classification.

With IVA elements in bulk, such as bulk carbon, etc., co-valency becomes apparent. In this group are lead and tin. These elements are usually thought of as having the metallic bond but this behaviour is not always true. Tin, for example, at very low temperatures, reverts back to the co-valent bond and will behave in a non-metallic manner.

Secondary bonds

van der Waals' forces

These forces act between *molecules* and not between atoms as is the case with primary bonds.

In many molecules, the centres of negative and positive charges do not coincide. In the water molecule, the co-valently shared electrons will tend to spend more time nearer the oxygen end of the molecule than near the hydrogen end simply because the oxygen nucleus carries more protons. Thus, one end of such a molecule carries a negative charge while the other carries a positive charge. The net result of these tiny dipole moments present on such molecules is that they attract each other weakly and this weak secondary bonding is an example of a van der Waals' force. Such bonds are easily broken by either mechanical or thermal stress.

ATOMS AND MOLECULES IN BULK MATERIALS

Organic materials

These are materials based on carbon and hence contain co-valent bonding. Of engineering importance are fuels and polymers.

Although the co-valent bond between the atoms in such molecules may be very strong, the bonds between different molecules are of the weak van der Waals' type and so bulk materials such as polymers tend to have low strengths and melting points.

The carbon atom has a co-valency of four. Compounds in which all four valency bonds are satisfied are referred to as *saturated* molecules. Conversely an *unsaturated* compound will contain multiple bonds between carbon atoms, since not all the available valency bonds are satisfied.

Formula of organic compounds

The valency bonds (electrons) are disposed tetrahedrally around the carbon nucleus and so an organic molecule is not strictly capable of being represented by a plane sketch. Thus in a long-chain molecule such as polyethylene, a plane sketch of formula would appear as

$$\begin{array}{ccccc} H & H & H & H & H \\ | & | & | & | & | \\ \sim C - & C - & C - & C - & C \sim \\ | & | & | & | & | \\ H & H & H & H & H \end{array}$$

but, in fact, because of the tetrahedral arrangement of the bonds, the molecule will look more like a helical coil with the backbone of carbon atoms taking up positions on a spiral, as shown in Fig. 1.6.

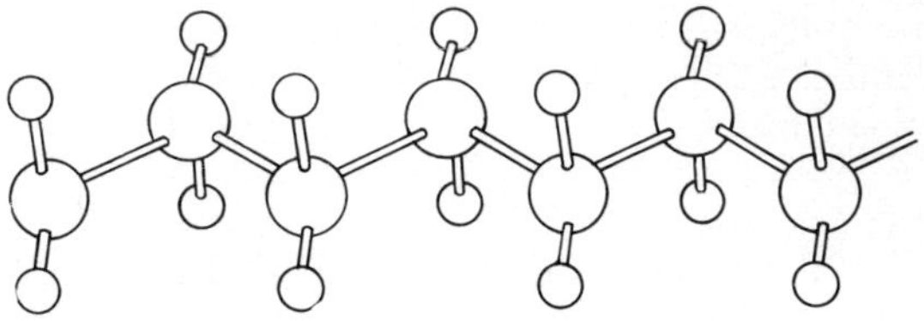

Fig. 1.6. Helical structure of polyethylene

Such a spring-like molecule is capable of uncoiling under stress and so many polymers are capable of exhibiting rubber-like behaviour.

It is, however, preferable to write organic formula as a structural sketch even though this is only a plane sketch. This is made necessary because of the phenomenon of *isomerism*. For example, the written formula C_2H_6O represents both ethyl alcohol and dimethyl ether, two entirely different compounds. The difference is at once evident if structural sketches are used:

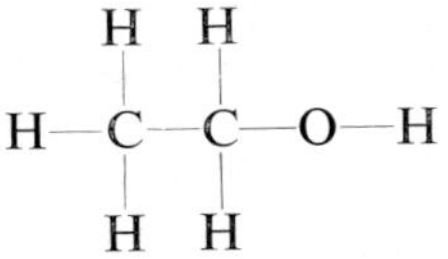

$$H\text{---}\underset{\underset{\displaystyle H}{|}}{\overset{\overset{\displaystyle H}{|}}{C}}\text{---}\underset{\underset{\displaystyle H}{|}}{\overset{\overset{\displaystyle H}{|}}{C}}\text{---}O\text{---}H \quad \text{ethyl alcohol}$$

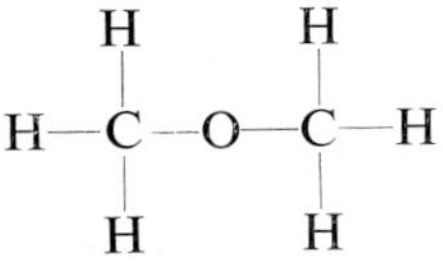

$$H\text{---}\underset{\underset{\displaystyle H}{|}}{\overset{\overset{\displaystyle H}{|}}{C}}\text{---}O\text{---}\underset{\underset{\displaystyle H}{|}}{\overset{\overset{\displaystyle H}{|}}{C}}\text{---}H \quad \text{dimethyl ether}$$

The two compounds are referred to as being isomers of each other. They are, of course, saturated compounds since each atom has its normal number of bonds. The bonds in such sketches are represented by the lines joining the atoms, a single line representing one co-valent linkage, i.e. representing one shared pair of electrons.

Classification of organic compounds

There are about one million known organic compounds, and so some sort of classification is necessary. As a preliminary, organic molecules may be classed as either *aliphatic* (compounds in which the carbon atoms are arranged in straight chains), or *aromatic* (compounds containing closed rings of carbon atoms).

Further classification can then be made into *homologous series*, the members of each series having similar properties.
Some typical examples of classifications can be considered.

Saturated hydrocarbons—aliphatic

These are hydrocarbons in which all four co-valencies of the carbon atoms are used. Because of this, they tend to have low chemical activity. Typical of saturated hydrocarbons are the members of the *paraffin homologous series* (Table 1.4).

Table 1.4. THE PARAFFIN HOMOLOGOUS SERIES

Name	Formula	Melting point K	Boiling point K	Specific gravity kg/m³	Form
Methane	CH_4	88	113	415·0	gas
Ethane	C_2H_6	101	180	446·0	gas
Propane	C_3H_8	—	228	536·0	gas
Butane	C_4H_{10}	138	274	600·0	gas
Pentane	C_5H_{12}	142	309	629·0	liquid
Hexane	C_6H_{14}	—	342	659·0	liquid
Heptane	C_7H_{16}	176	371	683·0	liquid
Octane	C_8H_{18}	217	399	702·0	liquid
Nonane	C_9H_{20}	-	423	718·0	liquid
Decane	$C_{10}H_{22}$	242	446	730·0	liquid

The series continues beyond this. For example, up to pentadecane $C_{15}H_{32}$, the compounds are liquids, while beyond this, waxy solids are produced.

The compounds are not very active chemically, because all bonds are satisfied. They are, however, useful as fuels since they readily oxidise:

$$CxHy + \left(x + \tfrac{1}{4}y\right)O_2 = xCO_2 + \tfrac{1}{2}yH_2O$$

Unsaturated hydrocarbons—aliphatic

Typical examples of this type of compound are contained in the *olefine homologous series* (Table 1.5).

In such compounds, the carbon valencies are not satisfied and so multiple bonds must occur.

Table 1.5. THE OLEFINE HOMOLOGOUS SERIES

Name	Formula
Ethylene	C_2H_4
Propylene	C_3H_6
Butylene	C_4H_8

Thus,   ethylene

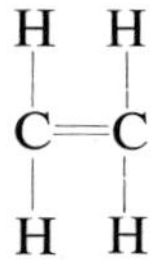

$$
\begin{array}{cc}
H & H \\
| & | \\
C & = C \\
| & | \\
H & H
\end{array}
$$

propylene

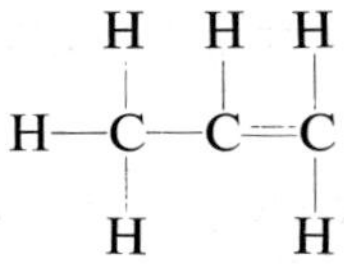

$$
\begin{array}{ccc}
 & H & H & H \\
 & | & | & | \\
H- & C- & C & =C \\
 & | & | & | \\
 & H & & H
\end{array}
$$

butylene

$$
\begin{array}{cccc}
 & H & H & H & H \\
 & | & | & | & | \\
H- & C- & C- & C & =C \\
 & | & | & & | \\
 & H & H & & H
\end{array}
$$

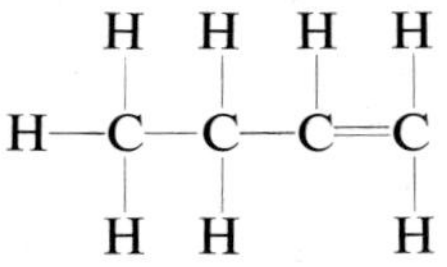

Because of the strains involved in multiple bonding, unsaturated compounds tend to be chemically active. When they do undergo chemical reactions, the reactions are additive, i.e. the multiple bonds are broken and another atom or group of atoms adds on to the free bond so created.

This sort of addition reaction can often be made to occur between similar molecules or *mers*. Hence, the single molecule or *monomer* of ethylene can be persuaded to open its double bond and add on to another ethylene molecule which has also had its double bond broken. This addition reaction may be repeated many thousands of times to give a long-chain molecule of the polymer polyethylene:

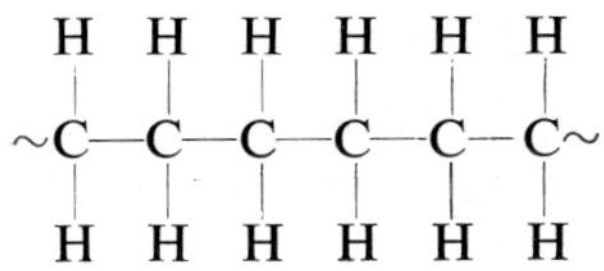

$$
\begin{array}{cccccc}
H & H & H & H & H & H \\
| & | & | & | & | & | \\
\sim C- & C- & C- & C- & C- & C\sim \\
| & | & | & | & | & | \\
H & H & H & H & H & H
\end{array}
$$

Similarly, the single molecule or monomer propylene can be induced to produce the polymer polypropylene:

$$
\begin{array}{c}
\text{H} \quad\quad \text{H} \quad\quad \text{H} \quad\quad \text{H} \quad\quad \text{H} \\
| \quad\quad | \quad\quad | \quad\quad | \quad\quad | \\
\sim\text{C}\!-\!\!-\!\text{C}\!-\!\!-\!\text{C}\!-\!\!-\!\text{C}\!-\!\!-\!\text{C}\sim \\
| \quad\quad | \quad\quad | \quad\quad | \quad\quad | \\
\text{H} \;\; \text{H}\!-\!\text{C}\!-\!\text{H} \;\; \text{H} \;\; \text{H}\!-\!\text{C}\!-\!\text{H} \;\; \text{H} \\
| \quad\quad\quad\quad\quad | \\
\text{H} \quad\quad\quad\quad \text{H}
\end{array}
$$

The aliphatic *aldehydes* are also unsaturated molecules, though they are not hydrocarbons. The two simplest members of this series are

formaldehyde

$$
\text{H}\!-\!\text{C}\!\!\begin{array}{c} \diagup\text{H} \\ \diagdown\text{O} \end{array}
$$

and acetaldehyde

$$
\begin{array}{c}
\text{H} \\
| \\
\text{H}\!-\!\text{C}\!-\!\text{C}\!\!\begin{array}{c}\diagup\text{H}\\\diagdown\text{O}\end{array} \\
| \\
\text{H}
\end{array}
$$

The aldehydes, particularly formaldehyde, are used in the manufacture of rigid thermosetting polymers such as phenol formaldehyde (bakelite) and urea formaldehyde.

Aromatic hydrocarbons

These are derivatives of the compound benzene C_6H_6. Such compounds must obviously contain multiple bonds since there are never enough hydrogen atoms to satisfy all the carbon valencies. However, in spite of this, the compounds show very little evidence of unsaturation and the multiple bonds are rarely broken during chemical reaction. So when aromatic hydrocarbons appear as part of the make-up of, say, a polymer molecule, they appear as whole, unchanged units on the side branches of the carbon backbone. They do not themselves open out to produce a chain.

The prototype of aromatic hydrocarbons is the benzene molecule C_6H_6:

$$
\begin{array}{c}
\text{H} \\
| \\
\text{C} \\
\diagup \;\; \diagdown \\
\text{H}\!-\!\text{C} \quad\quad \text{C}\!-\!\text{H} \\
| \quad\quad\quad || \\
\text{H}\!-\!\text{C} \quad\quad \text{C}\!-\!\text{H} \\
\diagdown \;\; \diagup \\
\text{C} \\
| \\
\text{H}
\end{array}
$$

From the point of view of usefulness as engineering materials, organic molecules are of interest in the form of polymers (plastics and rubbers). These materials are assuming ever-growing importance both in their own right and as replacements for conventional materials such as metals. The following tabulation gives repeat units (mers) of typical polymers:

vinyl chloride

$$\begin{matrix} H & H \\ | & | \\ C \!\!=\!\! C \\ | & | \\ H & Cl \end{matrix} \;\rightarrow\; PVC$$

tetrafluoroethylene

$$\begin{matrix} F & F \\ | & | \\ C \!\!=\!\! C \\ | & | \\ F & F \end{matrix} \;\rightarrow\; PTFE$$

adipic acid

$$\begin{matrix} O & H & H & H & H & O \\ \| & | & | & | & | & \| \\ H\!-\!O\!-\!C\!-\!C\!-\!C\!-\!C\!-\!C\!-\!C\!-\!O\!-\!H \\ & | & | & | & | \\ & H & H & H & H \end{matrix} \;\rightarrow\; nylon$$

Polymers such as polyethylene and nylon are partly crystalline inasmuch as they contain zones of long-range ordering of their atoms. Thus, if two chains of polyethylene are arranged

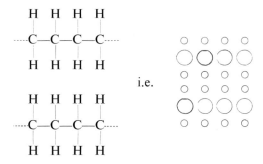

i.e.

then such an arrangement fulfils the definition of crystallinity. These crystalline zones are not continuous but are separated from each other by zones where there is no atomic ordering, i.e. amorphous zones. However, the existence of even partial crystallinity increases the rigidity of the material, particularly at elevated temperatures.

A polymer such as polymethylmethacrylate (Perspex) is completely amorphous. This is one reason why Perspex is optically transparent. Such polymers although they may be rigid at low temperatures become quite rubbery at elevated temperatures.

Inorganic materials

These are materials which are based on atoms other than the carbon atom and so include metals, alloys and ceramics.

One of the outstanding characteristics of metallic materials (and to a lesser extent of ceramics) is their crystallinity. When a metal or alloy freezes from the liquid condition, the atoms take up definite positions in space relative to each other. This regularity of atomic arrangement is termed long-range ordering and so the material is, by definition, a crystalline material.

Many of the important engineering properties of crystalline materials stem from the way in which the atoms are ordered within the crystal, and so to understand these properties it is necessary first to understand something of crystallography.

Elementary crystallography

Since atoms in metals are arranged in a regular three-dimensional pattern then there will be a simple unit which by repetition will produce the whole pattern. This is the *unit cell* and is the smallest assemblage of the packed atoms which bears the properties of the whole crystal. These unit cells are, of course, very small. For example, a cube of copper about 1×10^{-4} m edge length would contain about 22×10^{16} unit cells.

By joining up the centres of the atoms forming a unit cell, a geometric figure is outlined and it is on this geometric figure that the fundamentals of crystallography can be based.

Crystallographic notation

Most of the important processes which occur in crystalline material sinvolve movement of atoms within the crystal. When a metal is plastically deformed, for example, whole sheets of atoms move relative to each other in definite directions. It is thus necessary to be able to identify these planes of atoms and these directions, and a method has been evolved which is based on the reference axes of the unit cell, an example of which is shown in Fig. 1.7.

In a cubic unit cell, the reference axes would be X, Y, Z and distances along these axes are measured in terms of the unit cell side lengths a, b, c. The choice of origin O is arbitrary.

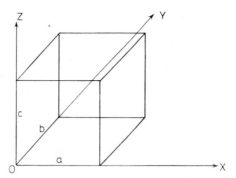

Fig. 1.7. Axes in a cubic unit cell

Using these axes, the *Miller indices* of any plane within the system can be defined by:

(*a*) selecting an origin which is outside the plane of interest;
(*b*) determining the intercepts of the plane on the reference axes in terms of distances *a*, *b*, *c*;
(*c*) taking the reciprocals of the resultant numbers;
(*d*) reducing the reciprocals to the smallest fractions which have a common denominator;
(*e*) omitting the denominator and enclosing the result in brackets.

Consider the three planes in Fig. 1.8 which intercept the axes *X*, *Y*, at 1, $\frac{1}{2}$, $\frac{1}{3}$ of their distances *a*, *b*.

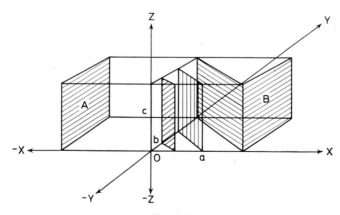

Fig. 1.8

3*

Taking the steps outlined previously:

	'1' plane	'$\frac{1}{2}$' plane	'$\frac{1}{3}$' plane
(a) origin	O	O	O
(b) intercepts	$1a,\ 1b,\ \infty c$	$\frac{1}{2}a,\ \frac{1}{2}b,\ \infty c$	$\frac{1}{3}a,\ \frac{1}{3}b,\ \infty c$
(c) reciprocals	$\dfrac{1}{1},\ \dfrac{1}{1},\ \dfrac{1}{\infty}$	$\dfrac{1}{\frac{1}{2}}\quad\dfrac{1}{\frac{1}{2}}\quad\dfrac{1}{\infty}$	$\dfrac{1}{\frac{1}{3}}\quad\dfrac{1}{\frac{1}{3}}\quad\dfrac{1}{\infty}$
	$= 110$	220	330
(d) reducing	$\frac{1}{1}\ \frac{1}{1}\ \frac{0}{1}$	$\frac{2}{2}\ \frac{2}{2}\ \frac{0}{2}$	$\frac{3}{3}\ \frac{3}{3}\ \frac{0}{3}$
(e) brackets	(110)	(110)	(110)

Thus, all planes having the same orientation in space will have the same indices.

Any plane which intercepts an axis on the negative side of the origin is indicated by a bar placed over the relevant number in the final indices. Thus, plane B would have indices (100), while plane A would have indices ($\bar{1}$00).

Location of the origin is arbitrary and by moving O by a distance $2a$ to the left in Fig. 1.8, both planes would become (100). There is therefore little point in distinguishing between planes which have the same indices ($h\ k\ l$) even though some may be negative. All such planes are members of a family. Miller indices, therefore, identify orientations of planes.

It is also usual to group together as families those planes which have a common property even though they may have different orientations. Such families are indicated by the type of brackets used. For example, the facial planes of a cube are (100), ($\bar{1}$00), (010), (0$\bar{1}$0), (001), (00$\bar{1}$), and these are designated as the {100} family.

Miller indices can also be used to designate directions in a crystal. Thus, in Fig. 1.9 the direction OP can be regarded as a vector passing through the origin.

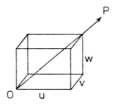

Fig. 1.9

The direction *OP* can be produced by travelling along *u, v, w,* and the distances moved are expressed in terms of *a, b, c.* The figures obtained are, as usual, reduced to the smallest whole numbers having the same ratio and the numbers are then enclosed in square brackets. Thus in Fig. 1.9 *OP* is reached by moving 1*a* along *u*, 1*b* along *v*, 1*c* along *w*, and so the indices of direction are [111]. Any other direction parallel to this will have the same indices. Negative directions are, as usual, designated by bars over the numbers and, again, directions with similar properties are grouped as a family and indicated by a difference in the type of brackets used. Thus a direction along an

$$X \text{ axis} = [100]$$
$$Y \text{ axis} = [010]$$
$$Z \text{ axis} = [001]$$

and these would be grouped as the $\langle 100 \rangle$ family.

In a hexagonal unit cell the position is complicated, since there are four axes. Three of these are co-planar at 120° to each other, while the fourth is at 90° to these. Fig. 1.10 illustrates this.

The planes are, as usual, specified in terms of intercepts with the four axes. Thus, basal planes are (0001), the front plane in Fig. 1.10 is (10$\bar{1}$0), while the side plane is (1$\bar{1}$00), as indicated.

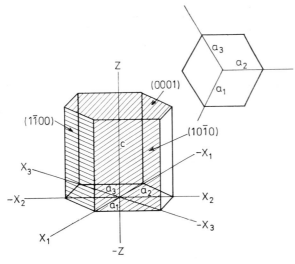

Fig. 1.10. Axes in a hexagonal unit cell

Directions in hexagonal systems are usually expressed as vectoral sums, as shown by the following example (Fig. 1.11).

A direction along a_1 has the same direction as the vector sum of three vectors: (*a*) a vector of length $+2a$ along the a_1 direction; (*b*) a vector of length $-1a$ along the a_2 direction; (*c*) a vector of length $-1a$ along the a_3 direction.

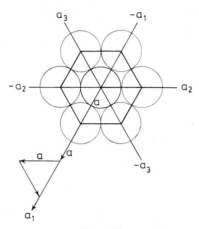

Fig. 1.11

There is no component along the *c* axis. The direction would thus be given as $[2\bar{1}\bar{1}0]$. Similarly, a direction along a_2 is $[\bar{1}2\bar{1}0]$ and along a_3 is $[\bar{1}\bar{1}20]$.

Crystals in metals

Most metals crystallise in one of a few simple arrangements. These typical arrangements are:

Close-packed hexagonal CPH

This is typical of the arrangement of the atoms in magnesium, zinc, beryllium, and some forms of nickel, chromium, titanium. It is also present in certain intermediate compounds formed in some alloys, for example $AgCd$, $CuZn_3$, Cu_3Sn, Cu_3Si, Mo_2C, Fe_2N.

In this structure, the basal layer of atoms is packed as closely as possible. The next layer is packed into the hollows of the first layer and the next layer is a repeat of the first layer. The stacking sequence of the atoms is thus *ABABAB...*, as indicated in Fig. 1.12.

The points marked on the unit cell diagram represent *only the centres* of atoms, and so the impression of openness conveyed by such diagrams is rather misleading.

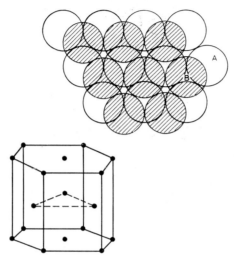

Fig. 1.12. Packing of atoms in CPH crystal structure

A metal crystal will, of course, be composed of many of these unit cells packed together and so the actual number of atoms per cell is not as great as appears at first sight. In each cell, each corner atom is shared by 6 cells and each face atom is shared by 2 cells, while the body atoms belong to one cell only.

The 'atom value' per cell is thus $(\frac{1}{6} \times 12) + (\frac{1}{2} \times 2) + 3 = 6$. Each atom in such an arrangement has 12 equally spaced neighbours. This is the co-ordination number of the structure.

Face-centred cubic FCC

This is typical of aluminium, copper, lead, gold, silver and of the compounds AlSb, Cu_2Mg. In this arrangement (Fig. 1.13) the first layer of atoms is close packed $-A$; the second layer lies over hollows in the first layer $-B$; the third layer lies over hollows in the second layer $-C$; the fourth layer is a repeat of layer 1 $-A$.

The arrangement is normally thought of as being cubic but if, say, an atom in layer *A* in the plan view is considered, it obviously has 12 near neighbours, i.e., 6 close neighbours in its own plane *A*; 3 close neighbours above in plane *B*; 3 close neighbours below in plane *C*.

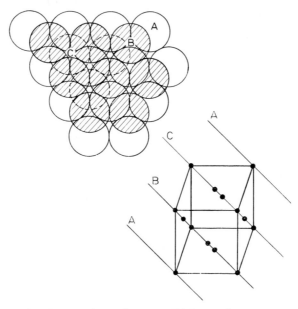

Fig. 1.13. Packing of atoms in FCC crystal structure

The co-ordination number is thus the same as in the CPH arrangement and so the FCC structure is simply a special case of CPH. This would mean, for example, that an FCC metal would be expected to alloy together with a CPH metal since the two structures can fit together.

The atom value of the FCC arrangement can be found as $(8 \times \frac{1}{8}) + (6 \times \frac{1}{2}) = 4$.

Body-centred cubic BCC

This is typical of iron, molybdenum, sodium and some forms of tungsten and chromium. It also occurs in the intermediate compounds AgZn, Cu_3Al, FeAl, Cu_5Sn.

The arrangement has a simple *ABAB* stacking sequence but both layers are open packed, as indicated in Fig. 1.14. The atom value

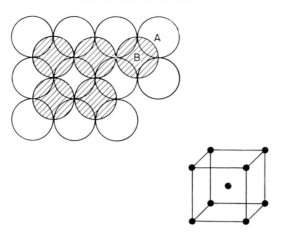

Fig. 1.14. Packing of atoms in BCC crystal structure

is $(8 \times \frac{1}{8}) + 1 = 2$ and the co-ordination number is 8. The arrangement is thus considerably more open in its packing than either CPH or FCC.

Body-centred tetragonal BCT

This arrangement is of interest since it is the arrangement of atoms found in quench-hardened steel. It is similar to the BCC arrangement except that the *c* axis is greater than the *a* axes (see Fig. 1.15).

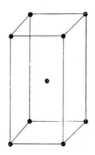

Fig. 1.15. BCT crystal structure

Polymorphism in crystal structures

Many metals are capable of having their atoms arranged in different patterns depending on the temperature. This phenomenon of polymorphism is often referred to as allotropy and is of importance because such allotropic changes involve volume changes and changes in mechanical properties. There is also the possibility of altering mechanical properties by rapidly cooling an allotropic material in order to suppress the change from the high temperature form to the low temperature form. The quenching and tempering of steel, for example, would not be possible but for the fact that iron is an allotropic material. The allotropic transformations in iron could therefore be considered.

Up to 1183 K, iron exists as BCC. This form or allotrope is called α-iron or ferrite and has an atom value of 2 atoms per unit cell. Between 1183 and 1676 K, iron exists in an FCC form called γ-iron or austenite, with 4 atoms per unit cell. From 1676 K to the melting point at 1808 K, iron exists again as BCC called δ-iron.

Thus, on heating through 1183 K BCC, iron having 2 atoms per cell will change to FCC iron having 4 atoms per cell and this closer packing of the atoms produces a contraction in volume of about 0·8%. These changes are reversible and act in the opposite way on cooling.

Changes in crystal structure, besides producing volume changes, also involve changes in internal energy and this appears as latent heat of transformation. Such changes therefore, at least in pure metals, occur at constant temperature.

If a pure metal exhibits allotropy then alloys based on that metal will do so as well. In an alloy, however, the transformation occurs over a range of temperature and not at a constant temperature.

Polymorphic changes involve quite drastic rearrangements of atoms and, since atomic movement in a solid metal is slow, it is expected that such changes would exhibit hysteresis, i.e. the structural change tends to lag behind the temperature change.

An example of this sluggishness of polymorphic changes occurs with tin. The theoretical transformation temperature is 286 K and a 25% volume change is involved. However, on cooling tin at normal rates the hysteresis is so pronounced that the transformation does not in fact occur until about 233 K. Hysteresis, in this case, is beneficial, since the large volume change associated with the change produces pronounced distortion and cracking.

The presence of hysteresis also gives the possibility of trapping a high temperature form down to room temperature by rapid cooling through the change. Rapid cooling may not be capable of completely stopping a transformation but it may allow only partial change. Thus, by quenching steel from its austenitic FCC condition, a very hard constituent called martensite is produced. This has a BCT crystal structure and is a sort of half-way stage between FCC austenite and BCC ferrite. Metals such as nickel, cobalt, titanium, manganese also exhibit allotropy.

Polycrystallinity in crystalline solids

Metals, alloys and many ceramics are fully crystalline in nature while some polymers are partially crystalline. The bulk material is not usually composed of a single crystal but of many thousands

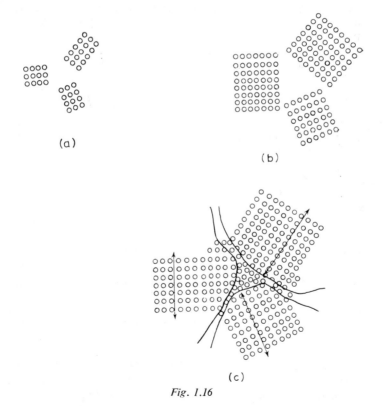

(a)

(b)

(c)

Fig. 1.16

of tiny separate crystals. The way in which this arises can be visualised by considering the solidification of a crystalline solid from the liquid condition.

As the liquid cools towards its freezing point, the atoms will lose kinetic energy, and at the freezing point the atomic bonds will be strong enough to hold small groups of atoms in the solid state pattern, latent heat being given out in the process.

This 'seed' production occurs at many points in the liquid mass and each seed crystal will then grow as more atoms take up positions on the pattern. Because of the random nature of seed production and growth, it is obvious that the growing crystals or grains will have different orientations. Fig. 1.16 illustrates three such growing crystals.

Where these differing orientations meet, a zone of mismatch will occur. This zone is only a few atom diameters in thickness but it has no regular atomic arrangement and could therefore be regarded as an amorphous or non-crystalline layer separating one volume of crystallinity from another. This amorphous layer is the grain boundary and its presence has a major influence on the properties of the bulk material since it has different properties from crystalline material.

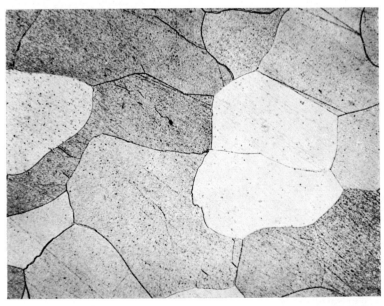

Fig. 1.17. Photomicrograph of grains and grain boundaries in metal

Bulk metal and some bulk ceramics are therefore polycrystalline in nature, as shown in Fig. 1.17.

The amount of grain boundary material can be altered simply by altering the grain size and this can be done by processes such as heat treatment or mechanical working. Such processes are thus capable of producing wide variations in mechanical properties of polycrystalline materials.

BIBLIOGRAPHY

BEYNON, C. E., *Physical Structure of Alloys*, Arnold, London (1946)
CARTMELL, E., *Chemistry for Engineers*, 2nd edn, Butterworths, London (1964)
FRANKEL, J. P., *Principles of the Properties of Materials*, McGraw-Hill (1957)
GORDON, M., *Structure and Properties of High Polymers* Plastics Institute, London (1957)
HUME-ROTHERY, W., *Atomic Theory for Students of Metallurgy*, London, Institute of Metals
HUME-ROTHERY, W., *Elements of Structural Metallurgy*, Monograph No 26, London, Institute of Metals
JASTRZEBSKI, Z. D., *Engineering Materials*, Wiley (1959)
Scientific American, Vol. 217 (Sept. 1967)
Structure and Properties of Materials (Ed. Wulff, J.), Vol. 1 'Structure' (1964); Vol. 4 'Electronic Properties' (1966), Wiley
TOTTLE, C. R., *Science of Engineering Materials*, Heinemann, London (1966)
VAN VLACK, L. H., *Elements of Materials Science*, Addison-Wesley (1964)

Chapter 2

ATOMIC MOVEMENTS IN CRYSTALLINE SOLIDS

Metals, some polymers and some ceramics are crystalline inasmuch as their atoms or ions are arranged in an orderly, repeating pattern which breaks down only at a grain boundary.

It is, however, illogical to expect that, in a crystal, every atom or ion will be in its appointed position on the lattice. Because of accidents of growth during solidification, the presence of foreign atoms of different size from the parent atoms, and so on, it is more logical to assume that faults will occur in the crystal lattice. This is, in any case, always true of a grain boundary.

The presence of irregularities in crystal structure is well established and they are vitally important since they control many of the bulk mechanical properties of the material and its response to thermal treatment. Some of these irregularities are described in the following pages.

POINT DEFECTS

These are defects caused by the absence of an atom from a lattice point or by the presence of a foreign atom. The result of such a defect is a local distortion of the normal lattice pattern.

A simple type of point defect is a *vacancy* caused by a missing atom, as shown in Fig. 2.1(a).

Because of the reduced strength of the bonding around the vacancy, the lattice will tend to collapse around this region. This sort of defect occurs in lattices composed of atoms or ions all of the same size, as with metals.

A special case of a vacancy which occurs in ceramics is the *Schottky defect*. In ceramic crystals, the lattice is built up from ions of different size and electric charge and, since overall neutrality must be maintained, the defect must involve pairs of ions, as shown in Fig. 2.1(b).

An interstitial point defect occurs in metals if one of the ions takes up an interstitial position instead of its normal position. This again will produce local lattice distortion, as indicated in Fig. 2.1(c).

In ceramics, the interstitial defect is the *Frenkel defect*, illustrated in Fig. 2.1(d).

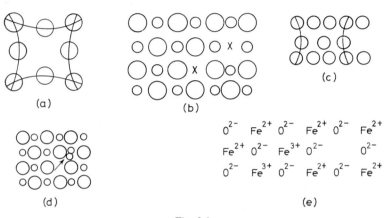

Fig. 2.1

Many ionic compounds amongst the ceramics are non-stoichiometric, i.e. the actual chemical composition is slightly different from that derived from the theoretical formula. Wustite is iron oxide and has a theoretical formula FeO. In fact, its actual chemical composition is more closely represented by the formula $Fe < {}_1O_1$. This arises because the natural material contains some ferric ions Fe^{3+} and, since electrical neutrality must be preserved, every two Fe^{3+} ions which are present must replace three Fe^{2+} ons and so a vacancy is generated in the lattice, as shown in Fig. 2.1(e).

These 'holes' in ceramic lattices have a major influence on the electrical characteristics and may induce semiconductivity.

Point defects are important in atomic movement processes such as diffusion. It is obvious that before an atom in a solid can change its position, it must have space into which to move. In a perfect lattice with no vacant spaces, such movements would not be possible. Heat treatment processes such as annealing, tempering, age hardening, surface hardening, all rely on movement of atoms in solids and so it is perhaps as well that lattices are imperfect.

LINE DEFECTS

These are defects in crystalline materials which extend for large distances through a crystal, often from one grain boundary to another. They are vitally important since they govern the bulk mechanical properties of the material. These line defects are known as dislocations and the idea of such defects in crystalline materials was first postulated, quite hypothetically, in order to explain the discrepancy between the calculated and observed strengths of metallic materials. However, since the first introduction of the hypothesis, about 1930, the existence of dislocations in real crystals has been adequately proved by electron microscopy.

The need to postulate some sort of major defect in a crystal lattice arose because of considerations of the following nature. A crystalline material such as a metal can be visualised as being composed of planes or layers of atoms, the orientation of which are describable by Miller indices (see Chapter 1).

If a stress is applied to a metal such that plastic deformation occurs, then atomic movement must have taken place within the lattice. Since, in a metal, such deformation occurs without fracture, then this means that the atoms must at all times have remained within 1 interatomic distance of each other. The movement must therefore involve a slipping or shearing of one plane of atoms over another. If it is assumed that the lattice is perfect, with every atom in its appointed position, then it should be possible to calculate the stress needed to cause this shearing of the planes.

Fig. 2.2 illustrates a portion of a perfect lattice.

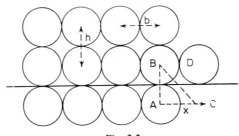

Fig. 2.2

Suppose ion *A* moves a distance *x* towards *C*. In doing so it will meet a resistance to its motion from, say, *B* until it is half-way between *B* and *D* and then will be in a stable position. Thereafter, it will experience an attraction towards *D*. The stress on *A* will

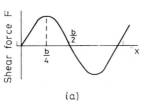

(a)   (b)

Fig. 2.3

therefore vary in a cyclic manner and the variation of required shear force will be as shown in Fig. 2.3(a), with the maximum force occurring at the $b/4$ position. It is logical to equate the onset of slip with the shear force at the $b/4$ value.

The shear strain at a displacement $b/4 = \theta$, as indicated in Fig. 2.3(b), but since for small angles θ approximates $\tan \theta$, this strain can be approximated to $(b/4)/h$.

Now, for a metal like aluminium, $h = b/\sqrt{3}$ and so the strain at the onset of slip should be

$$\frac{\dfrac{b}{4}}{\dfrac{b}{\sqrt{3}}} = \frac{\sqrt{3}}{4} = 0\cdot433$$

Since the shear modulus for aluminium $= 27\cdot6 \times 10^3$ MN/m², the theoretical shear stress needed to cause plastic deformation should be $27\cdot6 \times 10^3 \times 0\cdot433 = 11 \times 10^3$ MN/m². In fact, however, the observed shear stress needed to cause plastic flow in aluminium is $5\cdot52$ MN/m².

This vast discrepancy needs to be explained in some way and the inference is drawn that a real crystal lattice must contain many sources of mechanical weakness. These faults are the dislocations and they allow deformation to occur at remarkably low stresses. There are two main types of dislocation which can occur in crystalline materials, and these are described below.

EDGE DISLOCATIONS

In this case, the lattice is imperfect by reason of it containing an extra half-plane of atoms or ions running through some line in the crystal. Fig. 2.4 shows an edge dislocation and indicates that the dislocation is really a tunnel running through the crystal and that the lattice around the line of the dislocation is distorted.

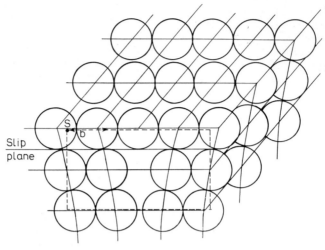

Fig. 2.4

If a shear stress is applied to the faulted lattice, the extra half-plane of atoms will move down the slip plane one interatomic spacing at a time leaving a more perfect lattice behind it. Fig. 2.5 illustrates this progress and it is seen that if the dislocation travels right to the surface of the crystal, it will produce a step on the surface.

This stepwise movement of the dislocation through the lattice requires that the shear stress need only break one line of atomic bonds at a time. In a perfect lattice, slip would only occur if the

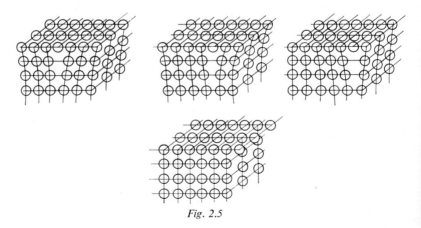

Fig. 2.5

total number of bonds over the whole slip plane were broken at one and the same time. Here lies the explanation of the discrepancy between calculated and observed strengths.

In plan view, an edge dislocation is the boundary between a slipped and an unslipped region of the crystal, as shown in Fig. 2.6. By definition, this boundary is normal to the direction of slip.

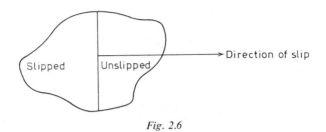

Fig. 2.6

The size of the dislocation is described by its *Burger's vector*. This is obtained by closing a circuit around the dislocation line. For example, referring to Fig. 2.4, if a circuit is produced by starting at *S*, moving 3 atomic units south, 4 east, 3 north, 4 west, then the circuit remains open by a distance equal to the dimension of the defect. A vector between the end of the circuit and the origin is the Burger's vector *b*. In this case $b = 1$ and the fault is 1 interatomic distance wide. The Burger's vector of an edge dislocation is always perpendicular to the line of the dislocation tunnel.

SCREW DISLOCATIONS

In this type of fault the irregularity of arrangement of the atoms, which constitutes the fault, occupies two separate planes of atoms perpendicular to each other, in a helical fashion.

The operation of a screw dislocation in plan view is illustrated in Fig. 2.7(a) and (b).

AB is the screw dislocation which moves towards *CD* as slip occurs from *B* to *A*. After a single step of slip the pattern would appear as in Fig. 2.7(b). Eventually, a slip step will appear on the surface of the crystal. A three-dimensional representation is shown in Fig. 2.8 and it is obvious here that the dislocation line is parallel to the direction of slip. The displacement of the half-crystals meeting on the slip plane is again the Burger's vector *b* of the dislocation.

4*

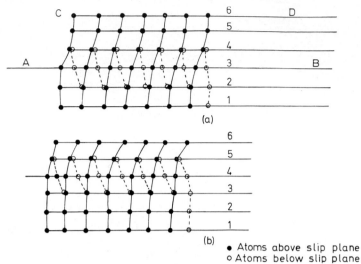

(a)

(b)

● Atoms above slip plane
○ Atoms below slip plane

Fig. 2.7

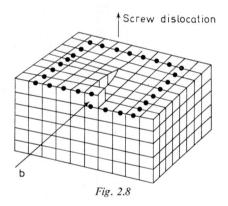

↑ Screw dislocation

b

Fig. 2.8

DISLOCATION LOOPS

The previous examples have shown both edge and screw dislocations as straight lines. In practice, the dislocation motion over a
slip plane tends to be in the form of an expanding loop or ripple—
rather in the same way as a ripple moves over the surface of a
pond. Such a loop is indicated in Fig. 2.9. It must be a combination of both edge and screw dislocations, the edge components
being perpendicular to the slip direction and the screw components
being parallel.

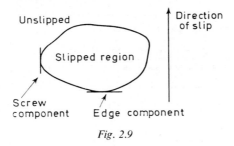

Fig. 2.9

GENERAL PROPERTIES OF DISLOCATIONS

Burger's vector

This describes the magnitude and direction of the dislocation move-ment. Because of the periodic nature of the force fields around the ions in a metal lattice, an ion must move from one equilibrium position to another, as was outlined in Fig. 2.3. The vector *b* will therefore always be in a direction connecting such equilibrium posi-tions and the direction can, as usual, be specified using Miller indices. The magnitude of *b* is the distance between equilibrium positions and very often, of course, this is one interatomic spacing. However, equilibrium positions can occur between the main lattice positions (see Fig. 2.3) and in such cases *b* is less than 1. Such 'imperfect' dislocations for which $b < 1$ produce a change in the stacking sequence of the planes of atoms, giving rise to stacking faults.

Intensity of dislocations

A normal metal crystal, no matter how carefully it has been pre-pared, will contain inherent or grown-in dislocations. A minimum intensity of about 10^{12} dislocation lines/m² of crystal surface is often quoted for a carefully grown crystal. In commercially pre-pared metals the inherent dislocation intensity will be higher than this.

Plastic deformation or cold work produces lattice distortions and actually creates more dislocations. In cold-worked metals, the dislocation intensity can be as high as 10^{15} lines/m² of crystal surface.

Generation of dislocations

When a metal is plastically deformed, it is observed experimentally that the slip process, at least in the initial stages, is concentrated on a few slip planes which are quite large distances apart. Fig. 2.10 illustrates the behaviour of aluminium.

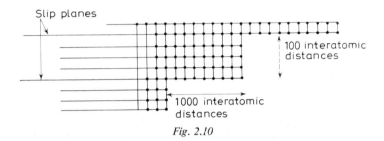

Fig. 2.10

The slip process will obviously produce steps on the surface of the grain, and these can be seen in Fig. 2.11. The fact that these steps are visible microscopically indicates that the active slip planes must be widely spaced—small increments of slip on adjacent planes would never be visible.

Fig. 2.11. Slip steps on the surfaces of plastically deformed crystals

Now a single dislocation moving down a slip plane can only produce a surface step of one atomic spacing in depth and there are not usually sufficient separate dislocations in a slip plane to account for the massive steps observed experimentally. This must mean that in some way a single dislocation must be capable of producing many units of slip and the Frank–Read source is usually quoted in explanation of this. The source is visualised as a dislocation line anchored at both ends (probably by grain boundaries). The application of a shear stress causes the line to bow, bulge and eventually regenerate itself so that a single source can create many dislocation loops or ripples. The process is illustrated in Fig. 2.12.

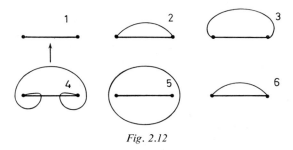

Fig. 2.12

Stress field around a dislocation

In an edge dislocation an extra half-plane of atoms is present on one side of a slip plane. This is analogous to a slit being made in a block of rubber and an extra sheet of rubber being inserted in the slit, and obviously such a situation involves a local strain. The extra half-plane will be under a compressive stress and so a corresponding tensile stress field will occur below the extra half-plane. A dislocation is therefore regarded as being surrounded by an elastic stress field rather in the same way that a magnetic pole is surrounded by a magnetic field.

Stress needed to move a dislocation

Both thermal and mechanical stresses can operate dislocations and, since only a few atomic bonds need to be broken, the dislocation will move, i.e. plastic deformation will occur, at remarkably low stresses.

Once dislocation motion has been initiated, it is observed that progressively more stress must be applied to maintain movement.

During plastic deformation, therefore, some agency must come into play which makes dislocation motion less easy.

One cause of hindrance is the increased density of dislocations caused by plastic deformation. Eventually, so many dislocations are generated that their stress fields begin to interact with each other and progressively higher stresses are needed to move dislocations through this 'traffic jam'. Any strain in the lattice will be surrounded by its own stress field and this stress field again will interact with the stress field around a dislocation and so jam it. Such lattice strains are produced by cold working, by solid solution formation, and so on.

The outline of the major properties of dislocations given above leads to some general conclusions regarding the behaviour of crystalline materials. It is evident that no real crystal lattice is perfect but contains many inherent or grown-in dislocations which can move through the lattice at stresses much lower that the calculated stresses. The movement of dislocations, which can only occur if a shear component of stress is present, produces slipping of one plane of atoms over another, i.e. plastic deformation. As long as dislocations can move, slip can occur but if movement is hindered, higher stresses must be applied to maintain motion. A crystalline material is therefore capable of work hardening, since the slip process itself creates more dislocations which then mutually interfere with each other's movement. One way of strengthening a metal, therefore, is by increasing dislocation density. The opposite is apparently also true since, if it were possible to produce a crystal with no dislocations, then that crystal would have the theoretical strength. The chance of dislocations being present in a lattice will decrease as the section size of the crystal decreases and, in the ultimate, a single row of atoms would contain no dislocations. Something approaching this condition can be produced by growing fine filaments or whiskers of crystalline materials having cross-sections as low as 10^{-12} m² and these, when tested, have extremely high strengths. For example, bulk iron has a tensile strength of only 250 MN/m² but iron whiskers have been produced with strengths up to 1200 MN/m². Such material is very fragile, has no ductility and is difficult to produce in long lengths and so present efforts are being directed to embedding the filaments in a soft, ductile matrix to produce composite materials (see Chapter 7).

Anything which interferes with dislocation motion in normal crystals will increase the stress needed to cause plastic deformation, i.e. will strengthen the material. As previously noted, foreign atoms in solid solution are capable of doing this and so we should expect alloys to be stronger than pure metals. A dislocation finds it impossible to penetrate through the jumbled arrangement of atoms which constitutes a grain boundary and, in fact, grain boundaries are very efficient in causing pile-up and jamming of dislocations. Hence, refining the grain size would be expected to strengthen a metallic material.

PLASTIC FLOW AND STRAIN HARDENING IN CRYSTALLINE MATERIALS

All crystalline materials contain dislocations, and plastic deformation will occur if these dislocations can be made to move.

The dislocations in crystalline ceramics can rarely be persuaded to move because of the presence of ions of unlike charge and in such materials the fracture stress is less than the stress needed to cause dislocation movement. Such materials, as a result, are brittle (see Chapter 6).

In metals, dislocations will usually operate at stresses below the fracture stress and so the following notes are mainly concerned with metallic materials.

Dislocation motion in metals first occurs at the elastic limiting stress and once it has occurred the metal is permanently deformed. The deformation is the result of the slip process taking place on certain crystallographic planes which move in definite directions.

The affected planes are the major slip planes which by definition are those planes on which the atomic population is highest. Reference to Chapter 1 would indicate that these planes are:

the   {110} family in BCC materials (6 per unit cell)

        {111} family in FCC materials (4 per unit cell)

and   {0001} family in CPH materials (1 per unit cell)

The major slip directions by definition are those directions of lowest atomic spacing, i.e.

$\langle 111 \rangle$ in BCC (4 per unit cell)

$\langle 110 \rangle$ in FCC (6 per unit cell)

$\langle 1120 \rangle$ in CPH (3 per unit cell)

These are not the only slip planes and directions which may be involved in plastic deformation. At elevated temperatures, further systems come into operation. For example, {100} planes become operative in FCC materials.

The stress needed to cause slip must be a shear stress and it is found that a certain critical value of stress is needed before slip will occur. Fig. 2.13 shows a piece of crystal with a slip plane orientated at an angle λ to an applied tensile force F.

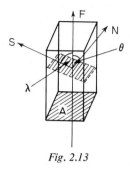

Fig. 2.13

Because of this orientation, F resolves into a true tensile component N and a shear component S. The area of the slip plane $= A/\cos\theta$ and so the stress on the slip plane $= (F\cos\theta)/A$ acting in direction F. Resolving this stress in the direction S gives

$$r = \frac{F\cos\theta \cdot \cos\lambda}{A}$$

and r is then the critical resolved shear stress. Before slip will occur, r must have some critical value dependent on the material, the plane being considered, temperature, and so on. r is lower for pure metals than it is for alloys and much lower for hot metal than for cold metal.

The expression for r indicates that the orientation of the slip plane will have a major influence, since the shear component of the applied stress will be greatest when $\theta = 45°$ and lowest when $\theta = 0°$ or $90°$. In a normal polycrystalline metal there are thousands of slip planes, the planes in one grain being orientated differently from the planes in neighbouring grains. Because of this polycrystallinity, an applied tensile stress will fall on planes whose orientations to the stress are randomly variable. It is evident, then, that on applying a stress above the elastic limit, slip or dislocation motion will occur first on those planes which are most favourably orientated to the applied stress. The greater the availability of slip planes and directions, the more deformation will be possible and hence we might expect FCC metals to be more ductile than, say, CPH.

Once slip has taken place on the most favourably orientated planes, dislocation multiplication occurs, and eventually dislocation motion on that plane stops because of jamming. Further slip must then occur on less favourably orientated planes and this, of course, requires a higher value of applied stress to give the required r value. In effect, the metal becomes capable of carrying higher stresses and this is the work or strain hardening effect. Towards the end of the process, the stresses being applied are so high that unfavourably orientated grains may be forcibly rotated into more favourable positions.

When all available slip systems have operated and become jammed, no further slip is possible and further application of stress will simply cause fracture because, by now, the stress S needed to continue slip becomes greater than the stress N needed to cause fracture.

The above observations on the response of a crystalline material to progressively increasing stress are reflected in the shape of the true stress–strain curve for a metal. Fig. 2.14 illustrates such a curve.

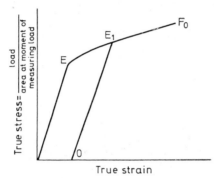

Fig. 2.14. Stress–strain curve for a metal

The rising part of the curve $E–F_0$ indicates that a progressively increasing stress is needed to maintain the slip process and reflects the progressive jamming of dislocations and the operation of progressively more unfavourable plane orientations.

Cold working is an important method of strengthening a metallic material since, as the unloading–loading curve $0–E_1–F_0$ indicates, the process results in an increased elastic limit or yield strength.

TWINNING

Crystals may undergo plastic deformation by a special form of slip called twinning. This mechanical twinning differs from ordinary slip in that:

(a) a change in orientation occurs between the slipped and un-slipped parts of the crystal;

(b) it involves slip on a number of adjacent planes;

(c) the amount of slip on one plane is directly related to the amount on neighbouring planes;

(d) the twin bands, unlike slip lines, are still visible on the surface of the crystal after polishing and etching because of the change in orientation produced.

Fig. 2.15(a) and (b) shows a twin in a crystal.

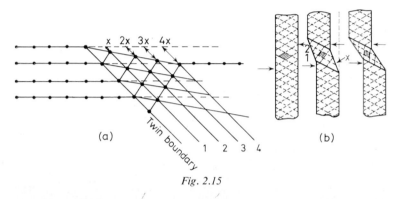

(a)

(b)

Fig. 2.15

Fig. 2.16. Photomicrograph of annealing twins (printed as a negative to accentuate grain boundaries and twins)

Plane 1 slips by, say, x interatomic distances. Planes 2, 3, 4, etc., will slip by some regular multiple of x. This organised slipping produces a difference in orientation across the twin boundary and so under a microscope a twin appears as a sharp-edged band running across the crystal face.

Mechanical twinning plays a minor part in the plastic deformation of BCC materials at normal temperatures but it is an important mechanism in CPH materials.

If a twinned crystal is reheated, the atoms, as in any deformed crystal, tend to rearrange themselves in a more regular pattern. The mechanical twins are not, however, wiped out since they often act as preferred sites for this reorganisation and so may be even more evident after annealing. Fig. 2.16 illustrates these annealing twins.

IMPORTANCE OF STRAIN HARDENING

Strain or work hardening increases yield strength and hardness and decreases ductility. This can be a nuisance in cold-forming operations such as cold rolling, since it means that progressively higher loads must be used to maintain the deformation. It also means that, at some stage, cold forming may have to be discontinued in order to anneal and soften the material so that further deformation can be carried out.

On the other hand, strain hardening is vitally important in increasing the strength of the final product. For example, cold rolling stainless steel sheet can more than double the yield strength, making possible important savings in weight and in materials cost. Again, many cold-shaping operations would be impossible without work hardening. In wire drawing, for instance, the reduction in section produces an increase in yield strength which more than compensates for the loss of cross-sectional area and so allows drawing to continue.

INFLUENCE OF PLASTIC DEFORMATION ON MICROSTRUCTURE

Changes in the external shape of a metal are reflected in changes in the shapes of the grains. If the deformation is carried out at temperatures below those which will allow diffusion to occur, the grains become progressively squashed down and elongated in the direction of principal flow. This change in microstructure with

Fig. 2.17. Photomicrograph of cold-worked metal

large amounts of deformation is shown in Fig. 2.17 and is accompanied by a loss of electrical conductivity and often also of corrosion resistance. The net result of cold working is, therefore, a banding of the microstructure and the production of fibre.

DISCONTINUOUS YIELDING AND STRAIN AGEING

Most materials exhibit a simple tensile stress–strain curve with only a change of slope between the elastic and plastic deformation regions. The yield point or elastic limit in such cases can be regarded as the lowest value of stress needed to cause dislocation motion.

BCC materials of good purity, however, behave oddly in that a *yield drop* is exhibited. The initial part of the tensile stress–strain curve for such materials shows the features illustrated in Fig. 2.18.

Between *O–E* the deformation is purely elastic and stress is proportional to strain. Between the elastic limit *E* and the upper yield point *UYP*, a sensitive testing machine will indicate a small amount of pre-yield plastic strain. When the *UYP* is reached, the stress suddenly relaxes to a lower yield point *LYP*. The material then plastically deforms at this reduced stress by an amount e_{LYP} without any work hardening. At the end of the yield point extension e_{LYP} work hardening sets in and the curve begins to rise.

A typical example of a material which exhibits such behaviour is low carbon plain carbon steel and most of the following notes

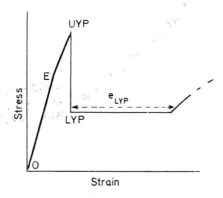

Fig. 2.18

concern this material since it is one of the most important available
to the engineer. It should, however, be remembered that similar
behaviour is exhibited by other BCC materials, such as niobium,
tantalum, molybdenum, etc.

In all cases, the yield drop is due to the presence in interstitial
solid solution of comparatively small atoms—usually carbon,
nitrogen or hydrogen. Of these, nitrogen is the main cause, at least
in mild steel.

A BCC lattice is an 'open' lattice since it contains only 2 atoms
per unit cell. Foreign atoms, provided they are small enough, are
capable of segregating into the inherent dislocations in the BCC
lattice. They will do this automatically since, by so doing, they will
expand the lattice to some extent and so relieve the stress field which
normally exists around a dislocation. The dislocation is thereby
made less of a fault than usual, i.e. it becomes more stable and, in
effect, is pinned or locked by the foreign atoms, so requiring higher
than usual stresses to move it.

This idea of dislocation pinning is illustrated in Fig. 2.19, which
shows a foreign atom positioned just below the extra half-sheet of

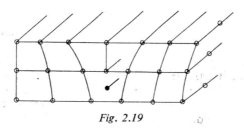

Fig. 2.19

atoms in an edge dislocation. There will be one such foreign atom per plane right down the dislocation tunnel.

An analogy which is often quoted here is that of a ruck in a carpet. The ruck (dislocation tunnel) can be imagined as running from one edge of the carpet (slip plane) to the other. It can be moved down the carpet very easily by applying shear force to push it along. If it is smoothed right out of the carpet, then the edge of the carpet (slip plane) at which the ruck (dislocation) disappears, will advance by a distance (slip) fixed by the original dimensions of the ruck. It is thus possible to move the carpet bit by bit at much lower stresses than would be required to move it bodily. However, if the original ruck (dislocation tunnel) is full of sand particles (nitrogen atoms) then it cannot be smoothed along at the same low stresses, i.e. the fault is pinned or locked.

In BCC materials, all the inherent dislocations can be regarded as being pinned by these 'atmospheres' of foreign atoms. These are the so-called Cottrell atmospheres.

The yield drop is sometimes explained in terms of these Cottrell atmospheres, i.e. the *UYP* is equated with the stress needed to tear the dislocations away from the pinning effect of foreign atoms. Once unpinned, the dislocations will continue to move at a much lower stress, the *LYP*. This is rather like the behaviour of a badly fitting door which needs a strong pull to start it to open but then needs a much lower force to continue the movement.

Considerable attention has been focused on the yielding behaviour of mild steel. It is of engineering importance, since the yield stress is the maximum working stress and mild steel, of course, is the major engineering material on a quantity basis. Any way in which the yield stress can be raised is therefore a valuable contribution to economy of engineering design.

The yield stress of first importance is, of course, the upper yield stress, since it is here that plastic flow first occurs on a large scale. However, the upper yield point is very difficult to measure since it is so sensitive to the presence of stress concentrations introduced by specimen geometry and finish, by the type of testing machine, and so on. In fact, with some testing machines and procedures, the upper yield point is never recorded. The lower yield stress is comparatively insensitive to testing conditions and so most design is based on this value.

The unpinning concept of the yield drop outlined above is attractive but it has certain disadvantages when the influence of grain size is taken into account. It is known that a fine-grained material gives a higher yield point than the same material in the coarse-grained condition. A plot of yield stress against the reciprocal of the square root of grain diameter usually produces a straight line, as indicated in Fig. 2.20. This method of analysing yielding, introduced by Petch, has proved useful in understanding the re-sponses of BCC materials to stress.

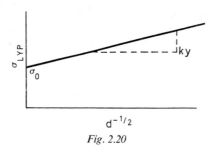

Fig. 2.20

The equation of the line is $\sigma_{LYP} = \sigma_0 + ky\,d^{-1/2}$ and it is at once apparent that the yield stress is composed of two separate stresses —the so-called Petch parameters.

The stress represented by σ_0 is the yield stress which would be exhibited by a single crystal ($d^{-1/2} = 0$) and so is not associated with the presence of grain boundaries. It is usually taken to be the stress necessary to move *free* dislocations through the lattice, i.e. a fric-tion stress associated with the lattice.

The stress $ky\,d^{-1/2}$ is obviously dependent on grain size and has been cited as the stress needed to unpin dislocations from Cottrell atmospheres. If it does represent an unpinning stress then it should be lowered by raising the temperature, since thermally induced vibration of atoms can reinforce applied mechanical stress. In fact, it has been clearly shown that $ky\,d^{-1/2}$ is not influenced by either temperature of testing or strain rate and so it appears unlikely that the yield drop involves unpinning of dislocations.

An alternative explanation of the yield drop makes use of the concept of dislocation multiplication. It has been shown that, in BCC materials, the inherent dislocations which are pinned by Cottrell atmospheres remain pinned, and take no part in the yield-ing process. What seems to happen is that during the pre-yield microstrain period *E–UYP* (Fig. 2.18), a few fresh dislocations are created from stress concentrations in the material. In BCC mat-erials in the plastic range, the value of stress carried by the material is very sensitive to dislocation velocity such that if the velocity decreases, the stress on the material also decreases and does so very sharply. It is therefore postulated that the first few fresh dislo-cations created during pre-yield microstrain undergo rapid multiplication. This larger number of dislocations must then, separately, move at lower velocities in order to give the strain rate which is being used and, hence, the stress drops also. This idea of the sudden creation of an avalanche of dislocations has some experimental verification and, of course, avoids the idea of dislocation unpinning. The stress $ky\,d^{-1/2}$ can now be regarded as the stress needed to create dislocations rather than to unpin existing dislocations.

The yield point extension e_{LYP} is usually considered as the period over which dislocation intensity is insufficient to give mutual interference. It does, however, represent plastic strain and, since more dislocations are created during plastic deformation, mutual interference will eventually set in and work hardening will begin.

Materials which exhibit a yield drop also exhibit other peculiarities such as a ductile–brittle transition (see Chapter 8) and are also prone to *strain ageing*.

Consider again the case of mild steel. Successive load–unload curves would appear as in Fig. 2.21. Curve A is the normal curve. The sample is strained just beyond e_{LYP} to, say, X, unloaded, and at once reloaded. This produces curve B and it is evident that the yield drop no longer occurs, the stress–strain curve simply retracing the path which it would have followed had the test not been interrupted.

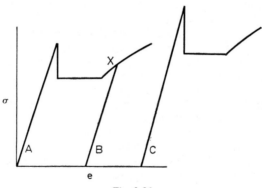

Fig. 2.21

Curve C represents a sample which has been strained to X, unloaded and then allowed to age or rest before reloading. The ageing may occupy a long time at room temperature or a much shorter time at elevated temperature. On retesting, the yield drop reappears and at a higher level than before. The material has thus been strengthened by straining and ageing. This reappearance of the yield drop occurs gradually over a period of time.

Strain ageing behaviour can be satisfactorily examined using a Petch analysis, and typical results are illustrated in Fig. 2.22.

Curve 1 represents tests on the original material. After yielding, the samples were further stressed to a constant strain beyond e_{LYP}, and these flow stresses are plotted on curve 2. The samples were then fully aged and retested to measure the new values of yield point, and these are plotted in curve 3.

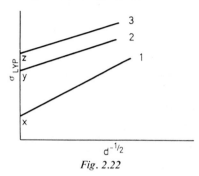

Fig. 2.22

It is obvious that the main effect of strain ageing is to raise the σ_0 part of the yield stress. σ_0 is the friction offered by the lattice to the movement of free dislocations and any agency which distorts the lattice will obviously raise σ_0.

After straining to beyond e_{LYP}, the samples will contain many more dislocations than they did originally, since plastic deformation creates dislocations. All the dislocations at this moment will be free (not associated with Cottrell atmospheres) and, hence, on immediate reloading no yield drop can occur because (*a*) if one is using the unpinning idea of yielding, a yield drop can only occur by sudden unpinning of dislocations; (*b*) if one is using the multiplication concept, yielding can only occur by a sudden massive increase in dislocation numbers stemming from a few initial dislocations. Hence curve *B* in Fig. 2.21 is explained. On ageing the material, the dislocations are gradually repinned as foreign atoms (nitrogen) diffuse back into them. There will therefore be a return of the yield drop, as shown in curve *C* of Fig. 2.21.

The analysis of strain ageing represented in Fig. 2.22 indicates that the increase in the yield point is mainly due to an increase in the σ_0 term. The lift is composed of an increment $x-y$ due to strain hardening alone (lattice distortion), plus an increment $y-z$ which is the lift actually due to ageing. This is the result of the pinning of the extra dislocations created during previous straining.

This phenomenon of strain ageing can produce a major fault in

5*

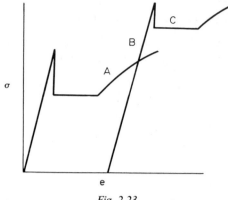

Fig. 2.23

pressed components of BCC materials. These faults are the unsightly surface markings called Lüders lines or stretcher strains. Presswork operations are carried out on cold-rolled sheet. This means that the sheet has been stressed beyond the yield point, say to point *A* in Fig. 2.23.

The application of this stress will produce dislocations in the material which are free from pinning by foreign atoms. If the sheet is pressed to shape immediately after rolling, pressing proceeds smoothly if a stress at *B* is applied, since no yield drop is present. If, however, the cold-rolled sheet is allowed to age before pressing, the dislocations will become repinned and the yield drop will progressively return at a higher level than the stress at which cold rolling was finished. The yield drop does not return homogeneously in all parts of the sheet and so if the same pressing stress *B* is now applied, those parts of the sheet in which the yield drop has not yet returned will deform since they will begin to flow at stress *A*, while those parts having a raised lower yield stress at *C* will not deform. This uneven flow produces ridges and wrinkles on the surface and these are the stretcher strain markings.

The effect can be prevented (*a*) if the sheet is pressed directly after rolling; (*b*) if the sheet is given a re-rolling treatment, just before pressing, to create fresh, unpinned dislocations; (*c*) if the foreign atoms responsible for repinning of dislocations are removed. In the case of mild steel, the main cause of the trouble is nitrogen and this can be removed by adding to the steel, at the ingot pouring stage, elements which will combine with any nitrogen present and precipitate it from solution. Aluminium or titanium are capable of doing this.

RECOVERY AND RECRYSTALLISATION

Cold work is defined as plastic deformation carried out at temperatures such that strain hardening persists.

If such a cold-worked material is heated to a sufficiently high temperature, its properties and microstructure will tend to revert back to the original condition. This change usually begins at a fairly well-defined temperature since a certain minimum activation energy must be supplied to trigger off the change. This temperature is the recrystallisation temperature or, more correctly, the recrystallisation temperature range, since the exact value varies with variation in external conditions.

In the reheating of a cold-worked material, the changes which occur do so in a number of distinct stages. A *recovery* stage may occur, this involving very little change in microstructure but producing a relief of internal stress concentrations. The recovery process constitutes an important stress-relieving operation applied to castings and welded fabrications.

The cold-worked material will be in a state of high internal energy, this energy being stored by the multiplication and mutual interaction of the dislocations. The dislocations will be piled up against obstacles such as grain boundaries and will be concentrated in the slip planes. During a recovery process, a redistribution of these dislocations occurs. This minimises the distortional energy associated with the pile-ups of dislocations but does not drastically reduce the total number of dislocations. Hence, internal stress concentrations are relieved but microstructure and overall mechanical properties are not greatly affected. In cubic materials, recovery occurs below the recrystallisation temperature and it is the subsequent recrystallisation process which produces major changes in structure and properties. In hexagonal metals, however, the process of recovery may go on to such an extent that the dislocations become substantially free once more and so major changes in properties can occur even without recrystallisation. The *recrystallisation* stage normally follows recovery and occurs at temperatures at which the atoms have enough mobility to rearrange themselves and reproduce the normal ordered arrangement pertaining before cold work was carried out. This, basically, is the production of new, equiaxed, strain-free grains out of the cold-worked structure; hence the name recrystallisation. The process involves the annihilation of dislocations and so produces dramatic changes in both microstructure and mechanical properties.

These changes on reheating cold-worked metal can now be looked at in more detail.

We can visualise a cold-worked material as a severely distorted structure with dislocations piled up and jammed at obstacles. Fig. 2.24 illustrates a section of a distorted lattice with dislocations piled up on the slip planes.

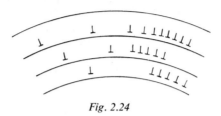

Fig. 2.24

During the initial heating, the recovery process occurs. This is basically a rearrangement of the dislocations and this can occur by dislocations climbing out of their own slip plane into other slip planes in which dislocation density is lower. Such a movement is a process of diffusion and can only occur if vacancies are present in the lattice. Fig. 2.25 illustrates the process of dislocation climb by vacancy diffusion.

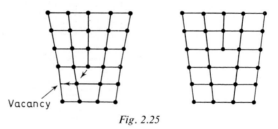

Fig. 2.25

By climbing out of a pile-up on one plane into another plane, the strain induced by the pile-up is reduced, i.e. stress relief occurs. This process of dislocation climb often produces an arrangement of regular 'walls' of dislocations, as illustrated in Fig. 2.26.

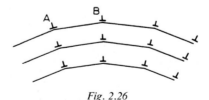

Fig. 2.26

These walls really represent low angle boundaries between slightly differently orientated parts of a crystal and the effect is called polygonisation.

On continued heating, or on raising the temperature further, the polygonised zones grow at the expense of one another, i.e. the dislocation wall *A* in Fig. 2.26 might diffuse towards *B*, leaving behind a larger section of dislocation-free crystal, as illustrated in Fig. 2.27.

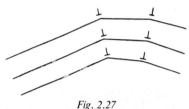

Fig. 2.27

Continuation of this process will, of course, increase the angle between the differently orientated sections.

It has been shown that the mobility of a boundary between sections of crystal of different orientations is a function of the boundary angle. The most mobile boundaries are apparently those of either very low or very high angle. On this basis, therefore, one would expect recovery to occur rapidly followed by some growth of the polygonised cells. This growth produces an increase in boundary angle, as shown in Fig. 2.27, and so growth would slow down. Eventually, after a period of slow growth, the boundaries will reach a critical angle and mobility will again increase so that the boundaries sweep through and out of the crystal, leaving larger, dislocation-free cells behind. This is really the process of recrystallisation, since the cells are really new, equiaxed grains.

The plot of grain size against temperature in Fig. 2.28 illustrates the stages in reheating cold-worked metal.

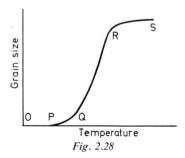

Temperature

Fig. 2.28

The period $O–P$ represents the incubation period during which dislocation rearrangement and polygonisation occur. There is then a slow start to recrystallisation as boundary angles build up to the critical angle, but when recrystallisation does start at Q, it goes forward rapidly over a narrow temperature range $Q–R$ and is then followed by a slow increase in grain size with continued heating.

The grain size of a metal governs its mechanical properties to a large extent and so, by controlling the process of recrystallisation, a wide variation in mechanical properties can be obtained. The process of recrystallisation is also important as a means of softening cold-shaped material so that further cold shaping can be carried out. The effects of recrystallisation on microstructure are shown in Figs 2.29 and 2.30.

Fig. 2.29. Early stages of recrystallisation—fine equiaxed grains (printed as a negative to accentuate grain boundaries)

The influence of cold work and recrystallisation on mechanical properties is indicated in Fig. 2.31. This actually refers to α-brass but the trend is similar for all metallic materials.

The ease with which recrystallisation occurs is a function of the internal energy of the cold-worked structure, i.e. of dislocation density and the severity of dislocation pile-ups. The higher this internal energy is, the greater is the driving force for change and so the more easily will the structure recrystallise on heating. It follows that the actual temperature at which recrystallisation occurs will not be a material constant but will depend on external factors.

Fig. 2.30. Grain growth stage of recrystallisation—large equiaxed grains exhibiting annealing twins (printed as a negative to accentuate grain boundaries)

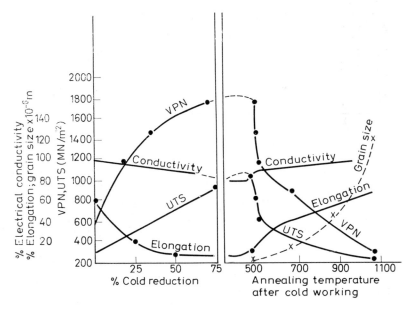

Fig. 2.31

In general, the material will recrystallise more easily, i.e. the recrystallisation temperature will be lower: (*a*) the greater the amount of prior cold work; (*b*) the finer the prior grain size; (*c*) the more pure the metal (since foreign atoms reduce the rate of diffusion); (*d*) the longer the time of annealing.

These factors are important when selecting a cold-shaped component for service at elevated temperature. Obviously, the service temperature must be well below that at which recrystallisation will occur. Table 2.1 indicates the recrystallisation temperatures of various materials.

In general, the recrystallisation temperature is related to the melting point. It is for this reason that materials for high temperature service (creep resistance) are based on the more refractory

Table 2.1. RECRYSTALLISATION TEMPERATURES OF VARIOUS MATERIALS

Metal	Recrystallisation temperature K	Melting point K
Lead Pb	below room temperature	600
Tin Sn	below room temperature	505
Zinc Zn	below room temperature	692
Copper Cu	373–473	1356
Aluminium Al	423–523	933
Silver Ag	473	1234
Iron Fe	673–723	1808
Nickel Ni	863	1727
Molybdenum Mo	1173	2893
Tungsten W	1423	3653

metals and, since alloying raises the recrystallisation temperature, alloys would be used here in preference to pure metals.

From this follows another definition of cold work, i.e. plastic deformation carried out below the recrystallisation temperature.

HOT WORKING

This could be defined as plastic deformation carried out above the recrystallisation temperature. As such, no permanent strain hardening should occur, since recrystallisation and softening will immediately follow the plastic deformation.

At hot-working temperatures, the yield stress is lower than

usual and more slip systems than usual are able to operate. Because of this, plastic deformation can be carried out at comparatively low stresses and so hot working is the favoured method for primary shaping of large masses of metal and for rapid production.

The function of hot working could be stated as: (*a*) to remove casting faults such as coring and segregation from ingots; (*b*) to weld up internal (clean-surfaced) cavities and blowholes; (*c*) to refine grain size; (*d*) to mass produce shapes cheaply; (*e*) to improve strength and toughness.

Most hot-working operations are deliberately finished at temperatures which produce a little permanent strain hardening in order to give a little better strength and to produce a final refined grain size.

The temperature range of hot working is between the recrystallisation and solidus temperatures of the material and so again is related to melting point. Table 2.2 indicates typical ranges.

Table 2.2. SOME TYPICAL TEMPERATURE RANGES OF HOT WORKING

Material	*Hot-working range* K
Steel	1573–1323
Nickel alloys	1450–1270
Copper alloys	1080–930
Aluminium alloys	780–680

The practice of finishing hot working just within the cold-working range gives some slight banding in the grains. A much more pronounced fibre occurs in alloys which contain precipitated phases (see Chapter 3). For example, in making brass, zinc is added to copper. Much of the zinc will dissolve in the solid copper, but if too much zinc is added the copper becomes saturated and the excess zinc is precipitated, not as pure zinc but as a chemical compound Cu_xZn_y. The final structure of such a high zinc brass would thus consist of grains of a solid solution of zinc in copper plus particles of precipitated zinc-rich compound.

In most alloys of this nature (two-phase alloys) the solid solution phase is soft and ductile while the precipitated compound is often very hard and brittle. This compound is also usually plastic at hot-working temperatures and so becomes elongated or filamented in the direction of working, so producing the characteristic fibre observed in many hot-worked components. The filaments of hard phase distributed throughout the soft phase tend to give extra stiffness and strength but it must be remembered that such properties are directional, i.e. the material will be anisotropic as regards mechanical properties. For example, the resistance to impact loading (toughness) would be high across the fibre but low if the stress was applied along the fibres. For these reasons, a hot-worked component should be shaped so that the fibre follows the contours of the component.

ATOMIC MOVEMENTS AND DIFFUSION

Diffusion is the movement of particles under the influence of some sort of driving force. Very often this driving force is a concentration gradient and the process of diffusion would eventually lead to equalisation of concentration. For example, in the freezing of an alloy of copper and nickel, there is a tendency for the higher freezing point nickel to solidify first and so form the grain cores, while the copper, which has a lower freezing point, tends to solidify later and so concentrates on the grain edges. This segregation or concentration gradient may be removed by the diffusion process if the casting is annealed.

Diffusion processes, as one would imagine, are rapid in gases and liquids but slow in solids.

Before any atom in a solid can move, it must have a vacant space into which to move. Hence, one begins to see the importance of structural imperfections such as vacancies in the process of diffusion in solids.

The activation energy which must be supplied to allow an atom to overcome bonding forces is not a constant value. For example, the smaller the diffusing atom the lower the activation energy. Similarly, atoms of materials having high melting point usually need high activation energies. The required energy is also higher if the diffusing atom must move via interstitial positions rather than via the normal lattice positions.

If the moving atom is the same as those in the bulk material, the process would be referred to as self- or homogeneous diffusion. Heterogeneous diffusion is the movement of one type of atom through another type of lattice and this is the process which is important in heat-treatment operations such as annealing, tempering and so on. The process of diffusion can be investigated quantitatively using *Fick's laws:*

(a)
$$J = -D\frac{dc}{dx}$$

where J is the mass of material moving across a unit surface area in unit time; c is the concentration; x is the distance, and so dc/dx is the concentration gradient; D is a diffusion coefficient and is negative, since the mass movement is in a down-gradient direction.

(b)
$$\frac{dc}{dt} = D\frac{d^2c}{dx^2}$$

This relates the concentration gradient to the change in concentration with time. Table 2.3 indicates some typical values of the diffusion coefficient D.

Table 2.3. SOME TYPICAL VALUES OF THE DIFFUSION COEFFICIENT D

Diffusing atom	Parent lattice	D at 773 K	D at 1273 K
Carbon	FCC iron	$10^{-10\cdot3}$ cm^2/s	$10^{-6\cdot5}$ cm^2/s
Nickel	FCC iron	10^{-19} cm^2/s	$10^{-11\cdot6}$ cm^2/s
Copper	Aluminium	$10^{-9\cdot3}$ cm^2/s	(liquid)
Zinc	Copper	$10^{-12\cdot2}$ cm^2/s	10^{-8} cm^2/s

It is indicated here that higher temperatures increase the diffusion rate D and that small atoms (C in Fe) will diffuse faster than large atoms (Ni in Fe).

The diffusion coefficient is a rate coefficient which is dependent on temperature. The Arrhenius relationship can often be used to express the rate at which a diffusion-controlled change will occur and its temperature dependency.

In its simplest form, the relationship is

$$\text{rate } R_x = \frac{1}{t} = Ae^{-E/kT}$$

where t is the time in seconds; A is a constant; E is the activation energy for the change, expressed in eV; T is the absolute temperature; k is Boltzmann's constant = $8\cdot614\times10^{-5}$ eV/K.
The constant k is often replaced by the gas constant $R = 8\cdot314\times$

10^3 J/kg mol K, in which case activation energy E would be expressed in J/kg mol.

In log form,

$$\log R_x = \log A - \frac{E}{2 \cdot 3kT}$$

and by plotting log R_x against $1/T$ a straight line results which has a slope equal to $E/2 \cdot 3k$. Hence it is possible to determine the energy needed to trigger off or activate a diffusion process. Conversely, if the other parameters are known, one can estimate rates of diffusion. The Arrhenius relationship is useful in following the changes which occur in any diffusion-controlled process in the solid state. It is used in later chapters to investigate such processes as age hardening, creep and heat treatment.

BIBLIOGRAPHY

DIETER, G. E., *Mechanical Metallurgy*, McGraw-Hill (1961)
'Diffusion in metals', *Scientific American*, Vol. 196, No. 5 (1957)
DOAN, G. E., *Principles of Physical Metallurgy*, McGraw-Hill
HONEYCOMBE, R. W. K., *Plastic Deformation of Metals*, Arnold, London (1968)
LEWIS, T. J., and SECKER, P. E., *Science of Materials*, Harrap, London (1966)
SMITH, M. C., *Principles of Physical Metallurgy*, Constable, London
The Structure of Metals, Institute of Metallurgists refresher course, London
    (1958)

Chapter 3

THERMAL EQUILIBRIUM DIAGRAMS AND THEIR APPLICATIONS

In a crystalline material at any temperature above absolute zero, the atoms are vibrating about their lattice points. The intensity of vibration increases with temperature, and the melting point can be regarded as the temperature at which thermal vibration is so intense that the bonds between atoms are broken. The atoms in the liquid are thus in a chaotic arrangement and possess both vibrational and kinetic energy. The way in which this liquid will solidify back into a crystalline solid will depend on its constitution and some typical cases can be considered.

FREEZING OF A PURE METAL

As the liquid approaches its freezing point, the atoms lose kinetic energy and, at the freezing point, the normal bonding forces between atoms become strong enough to establish small clusters or seeds of solid material in the liquid. The energy given out as the moving atoms settle down into the solid-state pattern appears as latent heat and this evolution counteracts the normal temperature drop. There is then a tendency for the temperature of the metal to remain constant until solidification is complete, and an ideal cooling curve would appear as in Fig 3.1.

A pure metal thus freezes and melts at the same constant temperature. A derived cooling curve is sometimes preferred. This is a plot of temperature against the time needed for the temperature to fall by a chosen amount. The derived cooling curve would appear as in Fig. 3.2.

The actual driving force for solidification is the degree of undercooling and all liquids must undercool to some extent before freezing will commence. Undercooling is the retention of liquid to

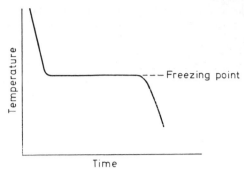

Fig. 3.1. Cooling curve for a pure metal

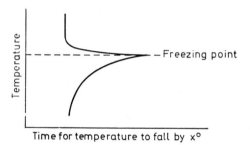

Fig. 3.2. Derived cooling curve for a pure metal

temperatures below the normal freezing point, and so when freezing does begin it tends to go forward more rapidly from more centres than usual. As a result, a highly undercooled or chill-cast liquid metal would be expected to have a finer grain size than usual.

During solidification, the atoms in a crystalline material take up positions on the solid-state pattern and the crystal grows. However, there may be a tendency for the crystal to grow more in some directions than in others. For example, in a cubic pattern heat losses would be greater from the corners of the cube than from the faces and so the cube would not continue to grow evenly in all directions. It often happens that a long spine of solid matter is first produced by growth in a preferential direction. This keeps on growing until it meets some obstruction such as the walls of the mould or another growing spine. If growth is restricted in one direction it continues in a direction roughly at right angles to the first by the production of smaller solid spines. This type of growth continues with the production of further branches of solid, again at right angles to previous branches, and the result is a bush-like growth which can be regarded as the skeleton of the grain, as shown in Fig. 3.3.

Eventually, of course, the skeleton is filled in as more metal solidifies. This freezing behaviour is called dendritic freezing, the typical tree-like structure being the dendrite. In a pure metal, all

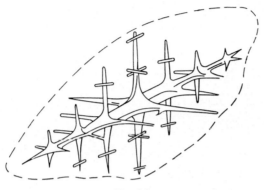

Fig. 3.3

parts of the dendrite are of the same composition and so would react similarly during preparation and etching for microscopic examination.

FREEZING OF IMPURE METALS OR OF ALLOYS

The mechanism of freezing is as before, but the behaviour is complicated by the presence of different components having different melting points.

In illustration, consider the freezing behaviour of an alloy of 40% of copper (freezing point 1356 K) and 60% of nickel (freezing point 1723 K). A cooling curve for this alloy would appear as in Fig. 3.4.

Temperature A marks the start of solidification and this temperature would be about 1643 K. Hence, it cannot be pure nickel which is freezing but will be a solid consisting mainly of nickel but carrying some copper, i.e. it will be a solid solution of copper in nickel.

Temperature B marks the end of freezing and will be about 1473 K. The solid produced will therefore consist mainly of copper but will carry some nickel, i.e. it will be a solid solution of nickel in copper.

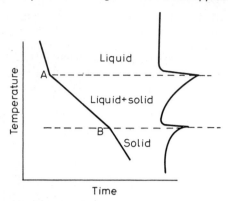

Fig. 3.4. Normal and derived cooling curves for an alloy

An alloy will therefore freeze and melt over a range of temperature in which liquid and solid exist together. This is the paste range of the alloy.

Dendritic crystallisation in such an alloy could be visualised as follows. At temperature A, the first solid seed crystals are formed. These grow in preferred directions giving the main spines of the

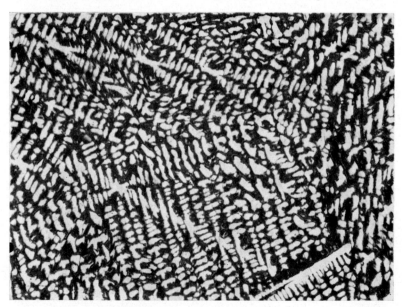

Fig. 3.5. Dendritic structure of cast brass

dendrites and these main spines will be rich in the higher melting point component. Between *A* and *B*, further solid dendrite branches will be produced and the solids being deposited will become progressively richer in the lower melting point component. At *B*, solidification is completed as the dendrite skeleton is filled in with solid which is rich in the lower melting point constituent.

A cross-section of a grain in which this sort of segregation has occurred would thus exhibit a core of material rich in the higher melting point component and such a crystal is said to be cored.

The dendritic crystal in this case is therefore heterogeneous and is visible on microscopic examination. Fig. 3.5 illustrates a dendritic microstructure.

The coring could be removed either by very slow freezing or by annealing after solidification, since then the differing compositions would be given an opportunity to merge or diffuse together to produce a homogeneous structure.

THE DIAGRAM OF THERMAL EQUILIBRIUM

The microstructure of a metallic alloy or a ceramic body has a large influence on engineering properties and it is therefore useful to be able to predict this structure at any given temperature. Even in a simple binary system, however, the number of possible alloys and temperatures of existence are infinite and so some 'blueprint' is needed which will conveniently summarise the required information. The equilibrium diagram serves this purpose and this can be plotted for a complete system from cooling curve results obtained on a few representative members of that system.

The information from which the diagram is plotted is obtained under equilibrium conditions, i.e. conditions of very slow heating and cooling, and so the diagram strictly refers only to such conditions. It cannot, for example, predict the structures obtained in, say, quenched steel. However, the diagrams will predict the structures produced on most normal commercial heating and cooling rates provided simple corrections are made.

Each alloy or ceramic system has its own phase diagram and so it is not possible to discuss every one. The following notes are therefore confined to a representative series of diagrams designed to establish the principles of their use. The phase diagrams for real materials are considered in subsequent chapters.

Phase or equilibrium diagrams are simply plots of temperature against composition. The composition axis is usually scaled in weight % but, in many cases, it is more convenient to use atomic %

which is, for a simple binary mixture of two elements X and Y,

$$\text{atomic } \% \; X = \frac{\dfrac{\text{wt } \% \; X}{\text{at. wt of } X}}{\dfrac{\text{wt } \% \; X}{\text{at. wt of } X} + \dfrac{\text{wt } \% \; Y}{\text{at. wt of } Y}} \times 100$$

BINARY SYSTEMS

Isomorphous systems

In this case, the two components are completely soluble in each other in all proportions and at all temperatures. This is analogous to, say, petrol–oil mixtures. Metallic alloys, such as cupronickels and gold–silver alloys, show this type of behaviour. In the liquid state the two components A, B dissolve completely in each other to produce a true solution which is maintained even when the liquid has solidified, and so a solid solution is produced.

A solution could conveniently be defined as a mixture on an atomic scale, and as long as this definition is fulfilled it is immaterial whether the solution is gas, liquid or solid. Thus in the solid state, atoms of B could (1), simply replace atoms of A in the normal crystal structure of A, giving rise to a substitutional solid solution which may be random as in Fig. 3.6(a) or ordered, i.e. the solute

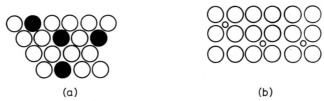

(a)   (b)

Fig. 3.6. (a) Substitutional solid solution: (b) interstitial solid solution

atoms substituted into the parent lattice in a definite pattern, or (2), if the atoms of B are much smaller than the atoms of A they may take up positions in the interstices between the A atoms, so producing an interstitial solid solution as in Fig. 3.6(b). This sort of solution is typical of the solution of carbon or nitrogen in steel.

Any homogeneous part of a system which has the same properties in every part of itself is referred to as a *phase*. A solution constitutes a separate or a single phase since the properties are the same in every part. Similarly, a system in which precipitation has occurred due to oversaturation would constitute a two-phase system.

In an isomorphous system there can be only one phase in the solid state, since the two components *A*, *B* are completely soluble in each other in the solid state. Solid phases which are either solutions or chemical combinations of the two pure components *A* and *B* are identified by using the Greek alphabet. Hence the solution of, say, copper in nickel would be referred to as an α solid solution since it is the first (and in this case the only) solid solution produced.

An isomorphous system could be investigated by preparing a series of mixtures of the pure components *A*, *B*, melting these to give liquid solutions and then plotting cooling curves as the solutions cool very slowly into the solid state. Such a set of cooling curves for the copper–nickel system would appear as in Fig. 3.7. By joining

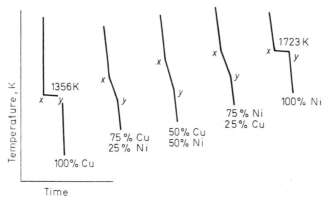

Fig. 3.7. *Cooling curves for the copper–nickel system*

together all *X* temperatures and, likewise, all *Y* temperatures the information can be replotted as an equilibrium diagram, as in Fig. 3.8. The diagram can now be used to predict the freezing behaviour of any alloy of copper and nickel. Consider the 60Ni : 40Cu alloy.

The line joining all *X* temperatures is referred to as the *liquidus* line of temperatures, since above this line single-phase liquid exists. The line joining all *Y* temperatures is the *solidus* line, since these are the temperatures at which solidification is complete.

The 60Ni : 40Cu alloy would cool without change until its particular liquidus temperature was reached at about 1640 K. At this temperature, assuming very slow cooling, the first solid crystals would precipitate from the liquid. This first solid, as indi-

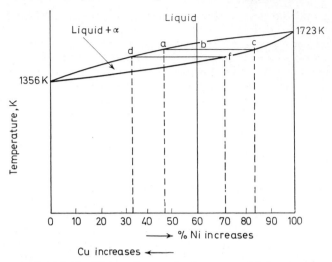

Fig. 3.8. Equilibrium diagram for the copper–nickel system

cated earlier, would be rich in nickel but would be a solid solution and so is designated α. Freezing would go on between the liquidus and solidus temperatures until, at the solidus, the alloy should be completely solid. Within the paste range, different compositions of α solid solution will have been precipitated, varying from a solid rich in nickel at the liquidus to a solid rich in copper at the solidus. Under equilibrium rates of cooling, all these different solid compositions will be given time to merge or diffuse into each other, so that by the time the alloy does reach the fully solid state, it should be composed of crystals all of which have the same composition, i.e. 60Ni : 40Cu.

The lever rule may be used to predict the condition of the alloy during freezing. A lever has a pivot and two ends. Thus a lever in the present context would be the line *a–b–c*. The average composition *b* is the pivot and the two ends connect together the two phases which exist together at this particular temperature. The lever or conode can only therefore be used in a two-phase region of the diagram and must refer to one temperature only, i.e. it must be horizontal.

At the temperature level *a–b–c*, the pot holding the alloy will contain some solid and some liquid. The solid will tend to be rich in nickel and so the remaining liquid will tend to be enriched in copper. Using the lever, the composition of the *solid* part of the paste is given by the intersection of the lever with the *solidus*, i.e.

point *c*. Projecting this on to the composition axis gives the composition of the solid α as 85Ni : 15Cu.

The composition of the *liquid* part of the paste is given by the intersection of the lever with the *liquidus*, i.e. point *a*. Projection on to the composition axis gives a composition of 45Ni : 55Cu.

The effect of nickel segregation into the solid is now clearly seen.

At a lower temperature *d–b–f*, the solid α will have diffused to composition *f*, while the remaining liquid will have reached composition *d*.

The solid α composition will vary down the solidus line until at the solidus temperature the whole alloy should be composed of 60Ni : 40Cu composition.

The lever may also be used to indicate the proportions of liquid and solid existing together in the paste range. At temperature level *a–b–c*, the pot holding the alloy contains some liquid of composition *a* and some solid of composition *c*. The point *a* represents the composition of all the liquid, while *c* represents the composition of all the solid. Hence the length of lever *a–b–c* could represent the total alloy and the partial lengths *a–b*, *b–c* will then represent the proportions of solid and liquid, respectively.

Therefore, at temperature *a–b–c*, the alloy consists of *a–b* parts of solid of composition *c*

$$= \frac{a-b}{a-c} \times 100\% \text{ of solid solution } c$$

and *b–c* parts of liquid of composition *a*

$$= \frac{b-c}{a-c} \times 100\% \text{ of liquid}$$

There is, therefore, an inverse relationship. The distances *a–b*, *b–c* may be measured using any scale but it is convenient, as usual, to use the composition scale on the diagram.

At temperature *d–b–f*, the proportion of solid will have increased to $(d-b)/(d-f) \times 100\%$, while at the solidus temperature the total length of a lever would represent the solid.

Once the temperature has fallen below the solidus temperature, there is no further microstructural change in the alloy and so the final condition should be a microstructure consisting of homogeneous grains of single-phase α solid solution.

Any other alloy within the system would freeze in exactly the same way and so all alloys exist as single-phase solid solutions at room temperature. These solid solutions are all α solid solutions since this is the first and only solid phase produced. They may differ as regards composition but microscopically they all appear to be the same, since they all exist in one phase.

Cooling an isomorphous alloy at rates greater than equilibrium will lead to coring in the crystals, as explained before.

The physical properties of an alloy in an isomorphous system tend to depend largely on composition. The two components will have atoms which differ in size, and so substituting an atom *B* into a lattice position normally occupied by *A* will cause either a local expansion or local contraction of the lattice depending on relative atomic size. Localised distortions of this nature are surrounded by elastic stress fields which interfere with the stress fields around moving dislocations. Maximum lattice distortion will occur at about the 50 at.% composition, and so in such systems many of the physical properties reach limiting values at around this composition.

In some isomorphous systems, the liquidus and solidus pass through a minimum temperature which lies below the melting points of both pure components. In such cases, the liquidus and solidus meet tangentially at the minimum point.

Some systems may show the opposite behaviour, i.e. the liquidus and solidus may coincide at a maximum. These cases are illustrated in Fig. 3.9.

The alloy of composition *X* behaves very much like a pure metal in that it melts and freezes at constant temperature. Such an alloy would not be prone to coring in its cast structure.

Systems which exhibit this behaviour include iron–vanadium, iron–nickel, copper–manganese and iron–chromium.

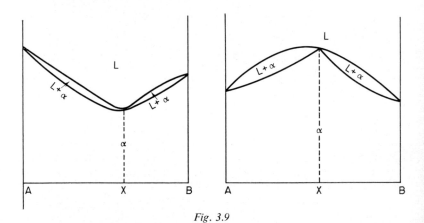

Fig. 3.9

Eutectic systems

In such systems the addition of either of the pure components to the other causes a lowering of the liquidus temperature. Two liquidus curves are then present which must meet at a temperature minimum at a fixed composition. At this fixed temperature and composition, any liquid present will freeze to give a characteristic structure called the eutectic. Some typical examples are:

The components are soluble in each other as liquids but completely insoluble in each other as solids

Tin and zinc behave in this manner and the equilibrium diagram could be built up by taking a series of cooling curves as shown in Fig. 3.10.

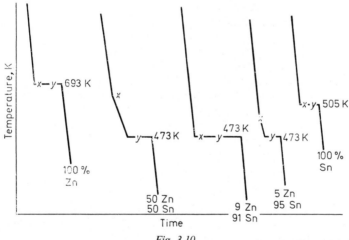

Fig. 3.10

The curves indicate that one particular alloy composition, the 91Sn : 9Zn behaves like a pure metal inasmuch as it freezes at a constant temperature. It is also apparent that other alloys in the system also finish their freezing process at the same temperature.

Replotting the data gives the equilibrium diagram shown in Fig. 3.11.

If the freezing behaviour of the 70Zn : 30Sn alloy is considered, it is seen that freezing begins at the liquidus temperature of 623 K.

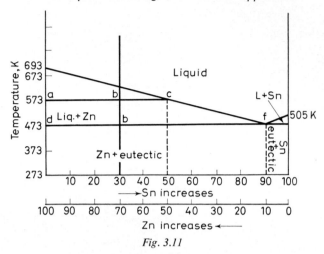

Fig. 3.11

The lever *a–b–c* indicates that the solid being deposited is pure Zn (composition *a*). This means that the remaining liquid is becoming richer in Sn and so, as temperature falls, the composition of the liquid will vary down the liquidus towards *f*.

At 473 K, the lever is *d–b–f* and the paste is composed of $(db/df) \times 100 = \frac{30}{90} \times 100 = 33\frac{1}{3}\%$ of liquid of composition $f(91\text{Sn} : 9\text{Zn})$ and $(bf/df) \times 100 = \frac{60}{90} \times 100 = 66\frac{2}{3}\%$ of pure zinc.

If this process of precipitation of pure zinc and enrichment of liquid in tin were to continue as cooling went on, the final liquid would end up as 100% tin and would freeze at 505 K. This is obviously impossible. According to the diagram, this, and any other alloy, finally freezes at 473 K. There is only one liquid composition which will freeze at 473 K and this is 91Sn : 9Zn. Thus, when the final liquid does freeze, it must freeze in such a way that the liquid composition does not change as the freezing process goes on. The final liquid in any Sn–Zn alloy solidifies by simultaneous precipitation of solid Sn and solid Zn in the ratio 91 : 9, so maintaining the temperature of freezing constant. The solid produced is the eutectic solid and frequently consists of plates of one metal interleaved with plates of the other.

Alloy compositions to the left of the eutectic composition will freeze to give pure zinc grains followed by eutectic, while compositions to the right will contain pure tin grains as the primary phase. In all alloys, the eutectic is always the same in that it is a laminated mixture of Sn and Zn in the ratio 91 : 9. The amount of eutectic will, however, differ in different alloys, as indicated in Fig. 3.12.

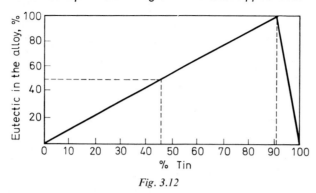

Fig. 3.12

The eutectic is the lowest freezing point constituent of such an alloy and, being the last to freeze, is liable to be found on the grain boundaries in the solidified alloy if it is present in only small quantities. The eutectic also tends to be stronger and more brittle than the primary phase and so the presence of a eutectic in an alloy may produce intercrystalline embrittlement. The eutectic will also be the first to soften and melt on heating and so may produce difficulties in the hot working of the alloy. Generally, therefore, alloys containing eutectics would be avoided in the manufacture of components which have to withstand high stresses in service.

The presence of eutectic may, however, be beneficial for cast components since the eutectic gives increased fluidity. Many die-casting alloys or alloys used for the production of complicated castings are either fully eutectic alloys or contain large quantities of eutectic material. Zinc-base die-casting alloys and cast irons are cases in point.

The components are soluble in each other as liquids and partially soluble in each other in the solid state

In this case component *A* can take some *B* into solid solution and vice versa. A saturation limit exists, however, and precipitation must occur at concentrations above this limit. This behaviour is typical of many engineering alloys, e.g. brasses, bronzes, aluminium alloys.

A typical equilibrium diagram is shown in Fig. 3.13.

The figure indicates the formation of a eutectic in all alloys between compositions *O* and *P*. The eutectic in this case is an intimate mixture of two solid solutions α and β. α is the solution

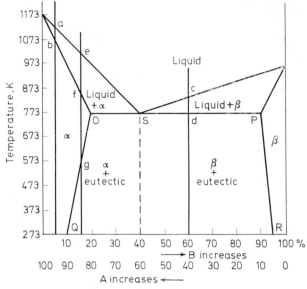

Fig. 3.13

rich in metal *A* and *β* is the solid solution rich in metal *B*. The sloping lines *O–Q*, *P–R* are the solid solubility boundaries. For example, at 773 K metal *A* is capable of holding in solid solution 20% of *B*. Solubility, however, decreases with temperature and so at 273 K, the α solid solution will contain only 10% of metal *B*.

Considering an alloy 95*A* : 5*B*, it is seen to begin to solidify at temperature *a* and is completely solid at temperature *b*. Since only 5% *B* is present, the resultant alloy consists of a single-phase solid solution α which is, of course, unsaturated. Any alloy containing 0–10% *B* behaves in a similar manner and such alloys behave as discussed under isomorphous systems. An alloy containing 60*B* : 40*A* undergoes a eutectic reaction. The liquid begins to freeze at *c*, and between the liquidus temperature *c* and solidus temperature *d* the usual paste exists consisting of grains of *β* solid solution mixed with liquid.

Use of the lever rule indicates that, at the eutectic temperature, the solid *β* grains have diffused to composition *P* (90*B* : 10*A*) while the liquid part of the paste has reached composition *S* (40*B* : 60*A*). There will be *Sd* parts of *β* (*P*) in equilibrium with *dP* parts of liquid (*S*). Any further fall in temperature will cause freezing of the eutectic liquid by simultaneous precipitation of solid *β* of composition *P* and solid α of composition *O*.

Assuming the alloy has just solidified at 773 K, its composition could be expressed as

1. $(Sd/Sp) = \frac{20}{50} \times 100 = 40\%$ of β of composition P
    and $(dP/SP) = \frac{30}{50} \times 100 = 60\%$ of eutectic composed of a mixture
    of α of composition O and β of composition P,

    i.e. $\alpha + (\alpha + \beta)$

or as

2. $(Od/Op) = \frac{40}{70} \times 100 = 57 \cdot 14\%$ of β of composition P
    and $(dP/Op) = \frac{30}{70} \times 100 = 42 \cdot 86\%$ of α of composition O
    i.e. $\alpha + \beta$

Both estimates really mean the same thing but the first case accounts for the distribution of the phases present, whereas the second case simply states the amounts of phases present.

The other possible behaviour is illustrated by alloys of composition between $O–Q$ or $P–R$. An alloy of composition $15B : 85A$ solidifies normally between temperatures e and f to give singlephase solid solution α. At temperatures between f and g the α is perfectly capable of holding the 15% of B in solid solution, but at temperature g the saturation limit is reached and a precipitate will be produced on further cooling. This precipitate cannot be a eutectic since this can only be produced from a liquid. Neither can the precipitate be pure B since solid solubility exists. The precipitate, as one would expect, is rich in metal B but also contains a little A, i.e. it is a precipitate of β solid solution.

This type of alloy is of interest in that the phase change which begins at g occurs in the solid state and so is liable to be sluggish. Rapid cooling of the alloy from about 673 K does not allow sufficient time for the precipitation process to occur and, in effect, the change which ought to occur at g does not take place. The rapidly cooled alloy may therefore finish up as single-phase α instead of two-phase $\alpha + \beta$. Since these two phases have different mechanical properties, suppression of β precipitation in this way may drastically alter the properties. The effect is also the basis of age-hardening processes discussed later.

Eutectoid systems

These are very similar to the previous eutectic systems except that all the transformations *occur in the solid state*. This could mean that one or more of the pure components is an allotropic material. A typical model system would appear as in Fig. 3.14.

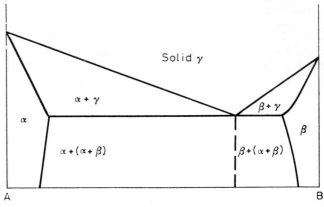

Fig. 3.14

The eutectoid is an intimate mixture of α and β and is produced by the decomposition of a single-phase solid solution γ. Apart from this difference, the diagram is operated in the same way as before.

These solid-state transformations are likely to be sluggish and so are often easily depressed or even suppressed by rapid cooling. The non-equilibrium structures produced are usually reasonably stable and may be extremely important, i.e. martensite in steel is the result of a suppressed eutectoid transformation and this phase is the basis of most high-strength engineering components made from steel.

Congruent transformations

When one phase changes directly into another without any change in composition, the phase change is said to be congruent. Certain intermediate compounds behave in this way. These compounds are often true chemical compounds having definite formula, and are formed by chemical reaction between the pure components at some specific temperature. Having a definite formula, such a compound will also have definite composition and so will appear at one specific point in the equilibrium diagram. Where the compound is not of true chemical formula, its composition will vary over a narrow range about an average composition. These intermediate compounds (or intermetallic compounds if the components are both metals) are very important in producing high-strength alloys. A compound called copper aluminide $CuAl_2$, for example, is responsible for the high strength of heat-treated aluminium alloys of the duralumin type.

The congruently melting intermediate phase will behave as a third component in the equilibrium diagram, i.e. it will melt and freeze at a constant temperature. A typical system is the Mg–Si system where the compound is centred around a composition Mg_2Si, as is shown in Fig. 3.15.

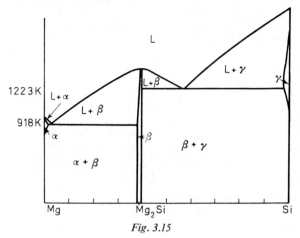

Fig. 3.15

The intermediate compound β effectively divides the diagram into two separate eutectic systems and so the freezing or melting behaviour of any alloy can be studied using the principles already outlined.

Peritectic transformations

In this type of system a transformation occurs such that a liquid and a solid solution react to produce a new solid solution. As always, the change is reversible, e.g. $L+\alpha \rightleftharpoons \beta$.

The peritectic transformation is evident in those alloy systems in which there is a wide difference in the freezing points of the pure components. Fig. 3.16 illustrates a typical peritectic system.

Consider the freezing of alloy X which passes through the peritectic composition P. Between the liquidus temperature a and the solidus temperature P, the liquid precipitates grains of α in the usual way. At P, solid α of composition O co-exists with a liquid of composition Q. The proportions will be $O–P$ parts of liquid Q and $Q–P$ parts of solid α (O).

The peritectic reaction now occurs between the liquid and the α. The two differing compositions O and Q diffuse into each other

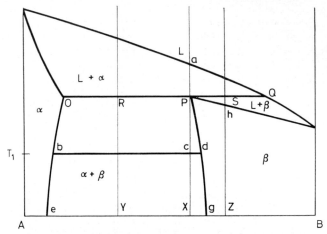

Fig. 3.16. A typical peritectic system

and in doing so solidification becomes complete with the formation of a solid of intermediate composition P. At the moment of
its formation, this new composition P is composed of the β phase
since it is assumed to lie on the triple-point junction on the diagram.

In the case of alloy X, the ratio of liquid Q and solid α (O) is
such as to allow the peritectic reaction to occur completely and so
100% β phase is produced.

On further cooling, the newly-formed β becomes saturated in
component A and begins to reject the α phase. Thus, at temperature T_1, the solid alloy will be composed of bc parts of β of composition d and cd parts of α of composition b.

At 273 K, the ratios will have changed to eX parts of β of composition g and Xg parts of α of composition e.

Alloy Y would again deposit α as a primary phase and would
again reach the peritectic temperature consisting of a paste of solid
α (O) and liquid Q. The proportions now, however, would be
$O–R$ parts of liquid Q and $Q–R$ parts of solid α (O).

There is, therefore, too much solid α and too little liquid Q for
the peritectic reaction to go to completion. The liquid will therefore react with *some* of the solid α to produce some of the new
phase β (P) leaving the excess α unused. The reaction will, as before,
occur in the ratio $O–P$ of liquid: PQ of solid α. After the peritectic
reaction, this alloy will thus be composed of solid α of composition O and solid peritectic β of composition P.

Alloy Z again reaches peritectic temperature as a paste, containing liquid of composition Q and solid α of composition O. The proportions are Q–S parts of α (O) and O–S parts of liquid Q.

In this case, there is too much liquid to produce a completely peritectic condition. Thus, some of the liquid reacts with all the solid α (O) in the correct ratio O–P : P–Q to produce some peritectic β (P). The alloy still contains liquid after the peritectic reaction is over and will, in fact, be composed of a paste containing P–S parts of liquid Q and S–Q parts of solid β (P). On further cooling between S and h, this remaining liquid freezes to give more β. The alloy finally solidifies to give a single-phase β condition.

Only those alloys between compositions O and Q will undergo the peritectic reaction. Alloys outside this composition range will solidify according to the isomorphous system.

Alloys undergoing a peritectic reaction will only do so properly with equilibrium rates of thermal change.

With normal cooling rates, the reaction is rarely complete since the reaction involves the slow interdiffusion of a liquid and a solid to produce a new solid. Thus, one could imagine a condition as shown in Fig. 3.17.

Fig. 3.17

Liquid has reacted with some α to give β which now forms a skin around the α and so separates the reactants. Such an enveloped structure often persists down to room temperature and, as a result, peritectic alloys may be badly cored and heterogeneous. This heterogeneity in microstructure, and hence heterogeneity in engineering properties, can be removed by annealing after casting.

GENERAL RELATIONSHIPS BETWEEN PHASE DIAGRAMS AND PROPERTIES

The equilibrium diagrams considered in this section are basic types designed to establish the principles of their use. It should be remembered that all the reactions which have been considered are reversible. For example, a eutectic change involves production of an intimate mixture of solids on freezing but, conversely, this solid

7

will melt at a constant temperature on heating to reproduce the eutectic liquid.

Strictly speaking, an equilibrium diagram should only be used to predict the condition of an alloy which has been very slowly heated or cooled. However, even with this limitation, the diagrams can supply valuable information and should be regarded as the blue-prints of alloy systems. Since the microstructure of an alloy governs its mechanical properties, the diagrams can be used to some extent to predict such properties.

Some typical examples of the use of the diagrams can be considered.

Alloys for cold working

Any alloy which is to be shaped by cold drawing, cold pressing, etc., needs to be soft, ductile and capable of absorbing large amounts of plastic deformation without becoming brittle. The phases which fulfil these conditions are usually the first phases in an alloy system, i.e. α phases. Second and subsequent phases are often considerably harder than the α phase and so their presence in an alloy would reduce the cold workability.

Many engineering alloys exhibit a phase diagram which includes the section shown in Fig. 3.18 and so it would appear that it is the dilute alloys which are amenable to cold shaping, e.g. low carbon steel, low zinc brass. A few alloy systems are isomorphous and so all alloys in such a system could be used for cold working.

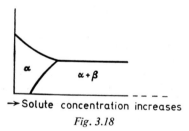

→ Solute concentration increases

Fig. 3.18

Single-phase alloys are therefore hardened and strengthened by cold working and could be recrystallised to give grain size control. This, in fact, is the only way of varying the properties of such alloys, since there is no possibility of quench hardening them. Rapidly cooling or quenching a single-phase alloy from below the solidus temperature produces no change of phase and so no change in mechanical properties.

Alloys for hot working

An alloy which is not readily shaped at room temperature must be hot shaped by rolling, forging, etc. Such alloys are usually too stiff and work harden too rapidly to be plastically deformed below the recrystallisation temperature. Usually, it is the presence of a second phase which gives this stiffening effect. Many of these second phases in alloy systems are intermediate compounds and are rather hard and brittle. The overall properties of a multi-phase alloy will thus depend on the relative proportions of soft tough α phase and hard, brittle second phase and it is obvious that the α phase should be in the majority. Hot working alloys therefore tend to be multi-phase alloys, e.g. $\alpha+\beta$-brass, medium carbon steels.

At the hot-working temperature, the second phase is often quite plastic and so becomes elongated in the direction of working. The final result is a banded structure consisting of hard filaments of second phase embedded in a softer α matrix. Such a structure will be anisotropic and will have different mechanical properties in different directions.

If the two-phase alloy contains a eutectic then care must be taken in predicting hot workability since the eutectic may tend to melt out at hot-working temperature and so produce hot shortness. The multi-phase alloys which are amenable to hot working are therefore those which involve a eutectoid or those in which the eutectic or peritectic temperature is well above the hot-working temperature.

Liquidus–solidus temperature range

The extent of the paste range varies with alloy type and alloy composition. As an alloy cools through its paste range, it deposits solids of continuously varying composition, e.g. compositions a, b, c, d, etc., as indicated in Fig. 3.19.

With very slow cooling, all these different compositions diffuse together to give the average composition X, but with normal cooling rates, diffusion is incomplete and the grains will be cored. Variability in chemical composition usually leads to variability in mechanical properties and corrosion resistance and so it may be necessary to anneal the cast alloy before putting it into service. The extent of the segregation and coring will depend to a large extent on the width of the paste range.

Rapidly cooling an alloy through its paste range aggravates the

7*

segregation effect. The solid which is deposited tends to be drastically over-enriched in the higher freezing point component. For example, in Fig. 3.20 the solid compositions deposited might be *R–S–T*, etc., instead of *O–P–Q*.

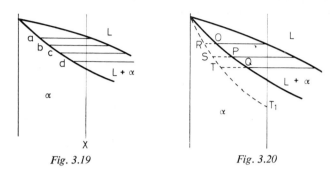

Fig. 3.19   Fig. 3.20

The result is a false solidus and so alloy X is not fully solid until temperature T_1 is reached. This may be important since it means that subsequent maximum annealing temperatures must be based on T_1 and not on the usual solidus temperature.

Alloys for castings

Components which do not have to carry high service stress are usually cast to shape since this is a cheap method of fabrication.

Liquid metal undergoes a shrinkage in volume of about 6% when it solidifies and unless this loss is made up by feeding the casting with more liquid metal, shrinkage porosity will result. It is also necessary for the material to have high fluidity so that it can flow easily to every part of a mould and take up an accurate impression of the mould.

These requirements are best fulfilled by using alloys which contain eutectics, the most fluid alloy being the fully eutectic alloy. Cast iron, zinc die-casting alloys, aluminium casting alloys are all examples of materials containing large amounts of eutectic.

Alloys responsive to rapid cooling

Such alloys must undergo a solid-state phase change. These changes involve the movement of atoms through solid material and so are likely to be sluggish. They can often be partially or completely suppressed simply by rapidly cooling the alloy through the state-change temperature. Alloys based on allotropic components are responsive to cooling rate and the special case of the quenching of steel is considered in detail in a later chapter.

A more general illustration of the importance of solid-state phase changes in alloys is provided by the phenomenon of age hardening.

PRECIPITATION FROM SOLID SOLUTION

Precipitation or age hardening as a means of improving the properties of alloys was first recognised by Wilm in the early 1900s. The process has always been particularly applicable to aluminium alloys but can be applied to almost any alloys, both ferrous and non-ferrous.

A necessary condition for precipitation is that a supersaturated solid solution should decompose into a multi-phase condition. The feature which is common to all alloys which undergo precipitation hardening is that their equilibrium diagrams exhibit a rapidly sloping solid solubility boundary, as illustrated by the line $X-Y$ in Fig. 3.21.

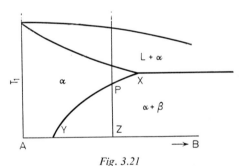

Fig. 3.21

An alloy Z in the slowly cooled state would consist of $\alpha+\beta$, the β probably being an intermediate compound, and so would introduce lack of ductility if coarsely distributed. The β would not, of course, be part of a eutectic.

The precipitation of β from α, which begins at the saturation limit P, is a solid-state change and as a result is sluggish. Such a change can be suppressed by rapid cooling. Rapidly cooling alloy Z, after soaking at T_1 to give single-phase α, would retain this condition down to room temperature. The α at room temperature would now hold in solid solution much more metal B than it ought to do, i.e. $Z-Y$ parts excess. It is, as a result, supersaturated and thermodynamically unstable.

On ageing, the α will tend to return to stability by rejecting the excess B in the form of β phase and it is this rejection which produces the dramatic change in mechanical properties. The rejection of β cannot be regarded as actual precipitation since there is no visible change in microstructure during the process. A precipitate becomes visible only after the maximum strengthening effect has been reached and so the appearance of a visible precipitate indicates an over-aged condition. Over-ageing involves a loss of strength and hardness and is to be avoided.

The driving force for rejection of β from supersaturated α will be related to the degree of supersaturation, but it is found that the supersaturation needed for initial nucleation of the rejected atoms is greater than that at which the particles of β will continue to grow once they have been formed. There is, therefore, a critical energy barrier which must be overcome before rejection will occur and this is the activation energy for the process. Nucleation of the rejected β occurs most easily at grain boundaries and at dislocations since these are both sites at which easy diffusion can occur.

Suppose, therefore, that an alloy has been solution treated, i.e. rapidly cooled to retain a supersaturated α condition, and that it is now to be aged. The general progress of the ageing or precipitation treatment is indicated in Fig. 3.22.

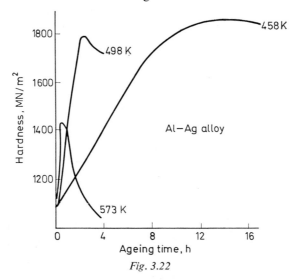

Fig. 3.22

In Al alloys, the mechanism of rejection of β is composed of at east two stages which depend on temperature: (*a*) 'cold' ageing, a reliminary stage occurring at low temperatures; (*b*) 'warm' ageing, subsequent stage occurring as ageing temperature is increased.

As the curves indicate, warm ageing produces quicker results than cold ageing but gives less pronounced hardening.

The cold ageing or preliminary stages correspond with the diffusion of solute atoms in the α lattice to produce clusters (Guinier–Preston or G.P. zones). These clusters are 10 to 100×10^{-10} m in size and about 100×10^{-10} m apart, and so there are very many of them in the α lattice. The cluster of *B* atoms will be coherent with the α lattice at this stage, i.e. will have the same crystal structure. The equilibrium precipitate of β will, however, probably have a crystal structure which is entirely different from that of the parent α lattice and so we could think of the cluster of *B* atoms as being a 'forced fit' in the α lattice. This forced fit introduces strain in the lattice and the stress fields around these clusters interfere with the stress fields of moving dislocations. The net result is that the formation of coherent clusters produces an increase in yield strength and hardness.

Warm ageing would correspond with the production of a transition state between the fully coherent β and the fully precipitated β. Lattice strains are not so pronounced and so the hardening effect is less.

Overheating, or too long an ageing time, will produce over-ageing. In this process, the transition lattice becomes unstable and actual precipitation of β particles occurs. Thus the β is removed from the α lattice, lattice strains are relieved and there is a loss of yield strength and hardness. For example, in an Al–Cu alloy (FCC structure), ageing might produce the following changes:

(*a*) *G–P* zone 1—a coherent cluster of Cu atoms in plate-like
formation about 50×10^{-10} m in size.
(*b*) *G–P* zone 2—a coherent cluster of Cu atoms having a
tetragonal structure.
(*c*) A transition structure only partially coherent with the
FCC α lattice
(*d*) A real precipitate of —$CuAl_2$— incoherent with the FCC
lattice and having an ordered tetragonal crystal
structure.

The appearance of $CuAl_2$ in the microstructure occurs after peak hardness has been reached and represents an over-aged condition.

Age hardening influences those mechanical properties which are sensitive to initial dislocation movement, i.e. yield strength and hardness. The maximum tensile strength is not so much affected. The creep-resistant properties of the alloy may, however, be improved since creep, being a process which involves dislocation motion, may be retarded by ageing.

An age-hardened alloy should not, of course, be operated close to the temperature at which it has been aged. Similarly, the weldability of such alloys is poor because feedback of heat from the weld deposit into the parent metal will produce a zone of over-ageing which results in a loss of strength, and may also give loss of corrosion resistance. Joints in age-hardened alloys are not, therefore, made by welding.

Recrystallisation will also remove the influence of age hardening and so any ageing treatment must be carried out below the re-crystallisation temperature. This is particularly important if the alloy has been cold shaped in between solution treatment and pre-cipitation treatment.

Age hardening involves atomic movement in the solid state and such reactions need a finite time to occur, since: (*a*) bonds between atoms must be broken; (*b*) atoms must diffuse to the reaction zone; (*c*) new phase interfaces must be formed when actual pre-cipitation occurs.

In the early stages of ageing, before definite precipitation has occurred, only the first two factors operate and the rate of change will follow an Arrhenius relationship. The relationship cannot be applied once true precipitation has started, since the extra energy needed to produce an interface between the precipitated phase and the parent phase considerably increases the reaction time.

Thus, ageing reactions in the initial stages are dependent on temperature via a relationship of the Arrhenius type:

$$R_1 = \frac{1}{t} = A\mathrm{e}^{-E/kT}$$

where R_1 is the reaction rate; t is the time in seconds; A is a constant; E is the activation energy; k is a constant (either Boltzmann's constant or the gas constant); T is the temperature in K.

In log form, the equation of a straight line results in

$$\log R_1 = \log A - \frac{E}{2 \cdot 3kT}$$

and so a plot of log $R \sqrt{1/T}$ should give a straight line of slope $-E/2 \cdot 3kT$, from which E can be derived. A knowledge of the activation energy E, which is assumed to be constant, will then allow the ageing rate to be derived for any given temperature.

Activation energies vary widely. Some alloys, for example, will age harden at room temperature and so these must be stored at sub-zero temperatures until the ageing effect is needed. Most commercial alloys, however, need to be reheated after solution treatment to cause ageing. Such alloys, according to the Arrhenius relationship, will theoretically age at room temperature, but the rate is so low that it can be ignored. The solution-treated super-saturated α would then be in a metastable condition, i.e. not thermo-dynamically stable but returning to stability at a negligible rate.

The ageing temperature chosen is usually that which will develop useful strengthening in reasonable time. The ageing can, of course, be speeded up by raising the temperature but, as the previous notes have indicated, this is liable to give lower final strength than if lower ageing temperatures over longer times are used.

BIBLIOGRAPHY

BRICK, R. M., and PHILLIPS, A., *Structure and Properties of Alloys*, McGraw-Hill
BURTON, M. S., *Applied Metallurgy for Engineers*, McGraw-Hill
COTTRELL, A. H., *An Introduction to Metallurgy*, London, Arnold (1968)
MARTIN, J. W., *Precipitation Hardening*, Pergamon (1968)
WINEGARD, W. C., *An Introduction to the Solidification of Metals*, Institute of Metals Monograph No 29 (1964)
WULFF, J., TAYLOR, H., SHALER, A., *Metallurgy for Engineers*, Wiley (1964)

Chapter 4

METALLIC MATERIALS

The remainder of this book is devoted to a review of the engineering properties of metals, alloys, polymers, ceramics and composites. It draws extensively on the unifying principles established in the previous chapters.

NON-FERROUS MATERIALS

Copper

Copper is one of the few metals which finds industrial use in the pure or nearly pure condition because of its high thermal and electrical conductivity. The various grades of copper used are:

(a) Electrolytic tough pitch copper (ETP), prepared by remelting electrolytic copper and casting into wire bars and billets preparatory to mechanical shaping. The remelting introduces about 0·02–0·05% oxygen, which gives the characteristic 'pitch' or toughness.

(b) Oxygen-free grades are produced by remelting electrolytic copper in an inert atmosphere to give better conductivity and ductility (oxygen-free high conductivity OFHC).

(c) De-oxidised coppers, produced usually by de-oxidising the liquid with phosphorus just before casting. A residual phosphorus content of about 0·05% occurs, resulting in reduced conductivity, but this material is easily welded (phosphorus de-oxidised copper PDC).

(d) Tough pitch grades produced by normal fire-refining methods. The residual impurity and oxygen content make the material unsuitable for conductivity work or welding but it is suitable for general purpose fabrication (TP copper).

(*e*) Arsenical coppers, containing up to 0·5% of arsenic in solid solution. Conductivity is poor but the material has a raised recrystallisation temperature and will maintain its strength up to 470–570 K. The arsenic also improves corrosion resistance.

Some typical properties of commercially pure coppers are given in Table 4.1.

Table 4.1

Element	*OFHC*	*ETP*	*TP*	*PDC*
Cu	99·95	99·9	99·5–99·85	99·85
Bi	0·001	0·001	0·003–0·02	0·003
Pb	0·005	0·005	0·01–0·1	0·01
Sb			0·005–0·05	0·005
As			0·05–0·1	0·05
Fe			0·01–0·03	0·03
O		0·03	0·1–0·15	

The mechanical properties (see Table 4.2) are not very different from one grade to the next.

Table 4.2

Condition	*0·1% Proof stress* MN/m^2	*Max stress* MN/m^2	*E* %	*VPN* MN/m^2
Cast	30·9	154	25	392·3
Hot worked	92·6	232	45	490·5
Recrystallised	62·0	216	50	441·5
Heavily cold worked	386	447·8	3	1226·3

Copper is a FCC material and as a result suffers very little loss of toughness at very low temperatures. The electrical conductivity of pure copper is one of its most important properties. Conductivity is drastically influenced by the presence of even small traces of impurities. Many of the impurities normally found in copper, such as arsenic, bismuth, oxygen, lead, phosphorus, form eutectic systems and may introduce hot shortness as well as reducing conductivity. The presence of oxygen may also induce gassing and embrittlement during welding.

Copper, with other elements, produces a very useful series of engineering alloys. Most of these alloys have the good corrosion resistance associated with copper. Like all non-ferrous materials, however, they tend to be expensive and so would only be used if they had some specific property to offer which is not offered by the cheaper ferrous materials. The properties of interest in copper base alloys are the conductivity and the corrosion resistance.

Copper–zinc alloys — the brasses

The brasses of industrial importance extend up to about 50% zinc and the equilibrium diagram for this range is given in Fig. 4.1.

The diagram indicates a wide range of primary α solid solution up to about 32% zinc at room temperature—this phase is FCC.

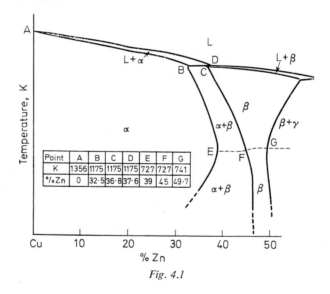

Point	A	B	C	D	E	F	G
K	1356	1175	1175	1175	727	727	741
%Zn	0	32·5	36·8	37·6	39	45	49·7

Fig. 4.1

Above about 32% zinc, the BCC β phase appears, this being based on the compound CuZn. The γ phase appearing above 50·6% zinc is of complex cubic structure and is based on Cu_5Zn_8.

Peritectic reactions occur at 1175 K and 1107 K, while ordering of the β phase occurs on cooling, as indicated by the dotted lines between 727 and 741 K.

The single-phase α-brasses are soft, tough and ductile and easily cold shaped. Their properties may be varied by cold work and recrystallisation but they cannot be heat treated by quenching. Recrystallisation is limited to temperatures below 870 K since, above this, rapid grain growth can occur producing poor surfaces (orange-peel effect) and difficulty in subsequent working.

α-brasses which have been severely cold worked are susceptible to a form of stress-induced corrosion known as season cracking. This is intercrystalline embrittlement resulting from preferential corrosive attack along the grain boundaries (see Chapter 8). The susceptibility to stress corrosion is reduced by stress-relief annealing after cold fabrication.

Some α-brasses which contain nickel and aluminium can be age hardened to give higher strengths than usual.

The two-phase α+β-brasses exist over the range 35–46·6% zinc. The β phase is based on the intermediate compound CuZn and is hard and brittle. α −β-alloys are thus more amenable to hot working than to cold fabrication. At hot-working temperatures, the β phase is plastic, and so pronounced fibre or banding is developed leading to anisotropy in mechanical properties.

There is a possibility, in these alloys, of retention of excess β by rapid cooling from the single-phase β region. Such an alloy would be harder and less ductile than usual.

In the cast condition the α −β-alloys tend to produce a Widmanstatten structure, i.e. on cooling from the β region into the α+β region, the β rejects α. This α precipitates preferentially on definite crystallographic planes giving a coarse structure in which the α is visible as sheets or plates, as indicated in Fig. 4.2.

The structure lacks ductility but may be removed by subsequent annealing or hot working. Season cracking may also occur in the α −β-brasses and often results from the residual stresses remaining after hot working, particularly if the working is finished at too low a temperature (see Chapter 8).

Single-phase β-brasses exist between 46·6 and 50·6% zinc. Being composed entirely of the compound CuZn, they tend to be very brittle at normal temperatures. Some use is made of the 50–50 alloy as brazing metal. The low copper content of the β-brasses results in poor corrosion resistance when compared with other types and these brasses are prone to intercrystalline attack from molten metals such as lead or tin and from solutions containing the ammonium radical. The straight binary β-brasses are not, therefore, of much use but by the addition of 5% aluminium, 1·5% iron, 1·5% manganese, a series of high tensile β-brasses is produced.

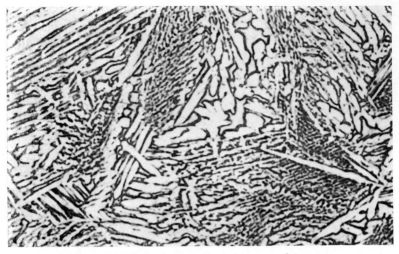

Fig. 4.2. Widmanstatten structure in cast α − β-brass

The mechanical properties of brass are, as usual, closely related to microstructure and particularly to the relative amounts of α and β present. This is illustrated in Fig. 4.3.

The graph indicates that ductility in α-brasses increases with zinc content up to about 30% zinc. The appearance of β causes a rapid loss of ductility, and with large amounts of β the embrittling effect also gives a reduction in tensile strength.

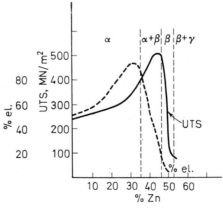

Fig. 4.3

Additions to brasses

Industrial brasses are not usually simple binary alloys. Small additions of other elements are usually made to confer specific properties. Typical of such additions are:

(a) *Lead*—This is insoluble in solid brass and so exists as minute globules in the microstructure. Its function is to give improved machinability and bearing properties.

(b) *Tin*—May be used to give added corrosion resistance, particularly in brasses for marine applications.

(c) *Aluminium*—Gives added corrosion resistance and, when added with tin, iron, manganese and nickel, produces a high-strength brass of $\alpha+\beta$ structure which can be used for non-sparking tools, gears, etc.

(d) *Silicon*—confers extra fluidity and better corrosion resistance. Silicon brasses may be used for die-casting.

(e) *Nickel*—when added in amounts between 10% and 30% to an alloy containing 55–63% copper produces the α-phase alloys known as nickel silvers. These alloys are readily cold-shaped and are used as a basis for silver plating and in the form of springs.

Copper–tin alloys — the tin bronzes

The phase diagram for this system is shown in Fig. 4.4 and has the following features:

(a) a peritectic at 1071 K at which $L_{(25.5\ Sn)}+\alpha_{(13.5\ Sn)} \rightleftharpoons \beta_{(22\ Sn)}$

(b) a eutectoid at 859 K at which $\beta_{(24.6\ Sn)} \rightleftharpoons \alpha_{(15.8\ Sn)}+\gamma_{(25.4\ Sn)}$

(c) a eutectoid at 793 K at which $\gamma_{(27\ Sn)} \rightleftharpoons \alpha_{(15.8\ Sn)}+\delta_{(32.4\ Sn)}$

The transformations below about 670 K are so sluggish that they do not normally occur and so commercial alloys are either α phase only or $\alpha+(\alpha+\delta)$ eutectoid. The limit of the single phase is very indistinct and depends largely on cooling rate, and so the diagram cannot be applied with any degree of certainty to commercial conditions.

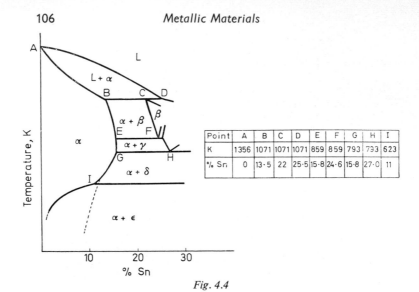

Point	A	B	C	D	E	F	G	H	I
K	1356	1071	1071	1071	859	859	793	733	623
% Sn	0	13·5	22	25·5	15·8	24·6	15·8	27·0	11

Fig. 4.4

The phases beyond α are based on intermetallic compositions. The δ phase, for example, is based on $Cu_{31}Sn_8$ and, as usual, tends to be hard and brittle, its presence tending to stiffen and strengthen the material.

The diagram exhibits a very wide $L+\alpha$ paste range, and coring and segregation during freezing are very pronounced. At normal rates of cooling, the α deposited in the paste range is much over-enriched in copper and the tin tends to concentrate into the remaining liquid. This effect is so pronounced that even in, say, a 5% tin bronze, the last liquid to freeze might well contain over 16% tin and such a liquid will freeze by a peritectic reaction to give β which will eventually produce δ in the final alloy. Cast tin bronzes thus tend to be fully dendritic and badly cored and even the very low tin bronzes will still contain the δ phase. This duplex structure is shown in Figs 4.5(a) and 4.5(b).

The α-phase boundary marked on the adjusted equilibrium diagram will therefore only apply if the alloy has been very slowly frozen or has been annealed after casting. Fig. 4.6 shows the structure of an annealed 5% tin bronze. Having been so treated, then single-phase alloys may be obtained with tin contents up to about 10% (i.e. ignoring the changes marked on the equilibrium diagram below 670 K).

The single-phase α-bronzes are tough and ductile and can readily be cold formed. Their properties may be varied by cold work and recrystallisation.

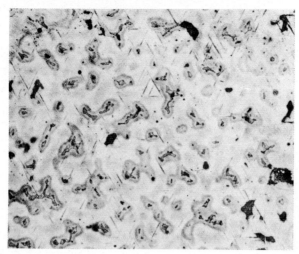

Fig. 4.5(a). Cast tin bronze. A cored dendritic structure with particles of the α + δ eutectoid in the darker-etching cores—the black areas are shrinkage porosity (× 100)

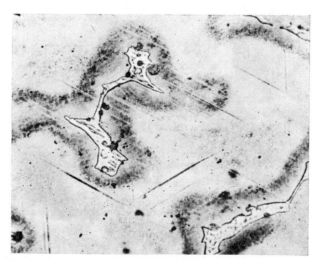

Fig. 4.5(b). Cast tin bronze. Photomicrograph of the cores in the α solid solution —the α + δ eutectoid is plainly evident in the centre of each core (× 500)

8

Fig. 4.6. 5% tin bronze annealed after casting, showing heavily twinned α grains
(× 100)

The majority of tin bronze is used in the cast form and such casting alloys usually contain the $\alpha + \delta$ eutectoid. The δ phase, if distributed evenly in the soft α matrix, confers good bearing properties on the alloy.

Phosphorus may be added to high tin bronzes up to about 0·5%. This produces hard, insoluble particles of copper phosphide Cu_3P in the microstructure and these improve the bearing properties.

If a bronze bearing is to operate under conditions of poor lubrication, lead may be added. Lead is insoluble in copper and will exist as globules in the microstructure. The lead provides lubrication by smearing out over the spinning shaft and, under severe conditions, it can maintain lubrication by melting.

Gun-metals contain a little zinc and are noted for their fluidity and castability. They have, like most tin bronzes, outstanding resistance to salt-water corrosion and so are widely used for marine castings.

Copper–aluminium alloys – aluminium bronzes

The equilibrium diagram, shown in Fig. 4.7, indicates: (*a*) a eutectic reaction at 1310 K, in which $L_{(8.5\ Al)} \rightleftharpoons \alpha_{(7.5\ Al)} + \beta_{(9.5\ Al)}$; (*b*) eutectoid at 838 K, in which $\beta_{(11.8\ Al)} \rightleftharpoons \alpha_{(9.4\ Al)} + \gamma_{2(15.6\ Al)}$.

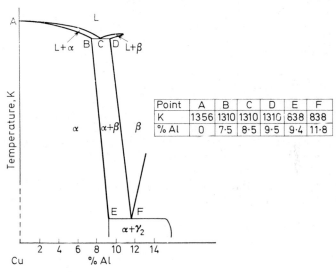

Point	A	B	C	D	E	F
K	1356	1310	1310	1310	838	838
% Al	0	7·5	8·5	9·5	9·4	11·8

Fig. 4.7

The α phase is a solid solution but β is based on the compound Cu_3Al, while γ_2 is based on Cu_9Al_4.

The eutectoid transformation at 838 K is very sluggish and so the γ_2 phase does not normally appear in commercial alloys.

The diagram also indicates that aluminium bronzes have rather high melting temperatures and this, combined with the very narrow paste range, points out the need for care in casting to avoid shrinkage porosity due to underfeeding.

The α and $\alpha+\beta$ phase field boundaries indicate an increase in solubility with decrease in temperature and so it is possible in alloys over 7·5%Al to retain excess β by increasing the cooling rate. The β phase is a hard intermetallic compound and so its presence will give increased strength but reduced ductility.

The single-phase α alloys existing up to about 9%Al are tough and ductile and are used for cold working. Like all aluminium bronzes, these alloys have outstanding corrosion and scaling resistance resulting from the impervious protective skin of Al_2O_3 which exists on the surface.

The duplex alloys containing $\alpha+\beta$ are of most importance. β is plastic at high temperature and so these alloys are widely used in the hot-worked condition. Hot working is carried out in the 750–1200 K range, producing the usual banded structure. Since solid-state phase changes occur in these duplex alloys, their properties may be varied by varying the cooling rate through these changes. Thus, consider an alloy 90Cu : 10Al:

(a) Equilibrium cooling rates would produce $\alpha+(\alpha+\gamma_2)$ eutectoid.
(b) Normal rates of cooling would give $\alpha+\beta$ only, since the eutectoid transformation is suppressed.
(c) Quenching from the single-phase β region produces an acicular or needle-like condition of high hardness. The high temperature β is BCC but the martensitic or acicular β, produced by rapid cooling, is CPH. This martensitic β can be reheated or tempered, at which it will tend to revert back to the final equilibrium condition. Thus the result of tempering is to produce a matrix of α phase with fine particles of γ_2 embedded in it. This 'sorbitic' structure is very tough.

The different properties which may be produced by varying the cooling rate are indicated in Table 4.3 which gives figures for 10% aluminium bronze. The table makes it obvious that the $(\alpha+\gamma_2)$ eutectoid should be absent.

Table 4.3

Treatment	UTS MN/m²	El %	VPN MN/m²	Izod J	Structure
Very slow cool	277·0	4	1372·0	1·36	$\alpha+(\alpha+\gamma_2)$
Hot worked—air cooled	509·5	28	1764·0	40·7	$\alpha+\beta$
Quenched from 900°C	663·9	4	2450·0	17·7	CPH β
Quenched and tempered	756·6	29	1862	40·7	$\alpha+$particles of γ_2

Although heat treatment is therefore a possibility it is not much used in practice, since equally good properties can be obtained in forged or cast products simply by adjusting the air-cooling conditions after fabrication. Figs 4.8(a), 4.8(b) and 4.8(c) illustrate the microstructures of aluminium bronzes in various conditions.

Fig. 4.8(a). Very slowly cooled duplex aluminium bronze, showing α phase (white) with the α + γ₂ eutectoid (courtesy the Copper Development Association)

Fig. 4.8(b). Hot worked duplex aluminium bronze, showing stringers of β phase (courtesy the Copper Development Association)

Fig. 4.8(c). Quenched duplex aluminium bronze, showing the acicular structure (courtesy the Copper Development Association)

The alloys in commercial use are usually restricted to a fairly narrow range of compositions, i.e. α alloys for cold working contain 4–7% Al, and duplex α+β alloys for hot working or casting contain 9–11% Al. The explanation for this is indicated in Fig. 4.9.

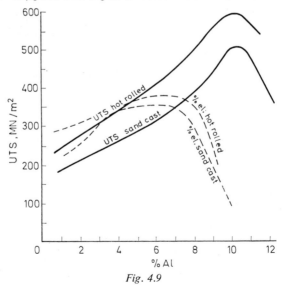

Fig. 4.9

Additions which may be made to aluminium bronzes include:

(a) Lead—to give improved machining properties. Lead in copper-base alloys reduces ductility and so leaded alloys cannot usually be hot forged or rolled but must be fabricated by extrusion.

(b) Manganese—is used in small quantities for deoxidation but in amounts over 5% a series of magnetic alloys, the Heusler alloys, is produced.

(c) Iron—is often added as a grain-refining element. Aluminium bronzes undergo rapid grain growth at high temperatures and thus iron is added to retard this grain growth during heat treatment.

(d) Nickel—may be present to induce age-hardening characteristics.

Copper-nickel alloys—The cupronickels

The phase diagram for the system is given in Fig. 4.10. The diagram is of the simple isomorphous type. Some of the nickel-rich alloys are magnetic and the Curie temperature varies linearly from 626 K for pure Ni to 103 K for 50Cu: 50Ni.

All alloys are single-phase α and exhibit pronounced coring in the cast state. Annealing may thus be necessary before cold fabrication. Apart from their good cold workability, the alloys also have

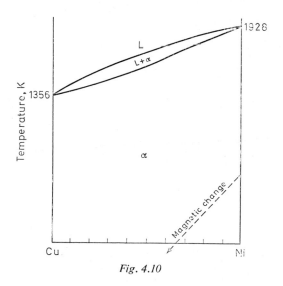

Fig. 4.10

outstanding corrosion resistance which increases with nickel content.

Alloys containing nickel are susceptible to attack by atmospheres containing sulphur and an embrittling NiS eutectic is produced. Carbon pick-up can also cause loss of ductility. These alloys are therefore best melted in electric furnaces.

The properties of cupronickels vary directly with the nickel content as shown in Fig. 4.11. Electrical resistance reaches a maximum at about 50Cu: 50Ni and alloys of this type are used for resistors, particularly for precision instruments.

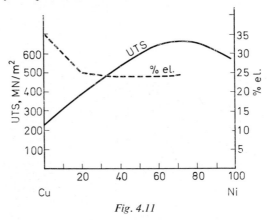

Fig. 4.11

Addition of up to 30% zinc to Cu–Ni alloys produces the nickel silvers, some of which have higher elastic modulus than is usual with copper-base alloys and so find application in spring manufacture.

Addition of Si, Al or Fe to cupronickels induces age-hardening characteristics and high-strength alloys can be produced in this way.

Copper–beryllium alloys—beryllium bronzes

The phase diagram, given in Fig. 4.12, exhibits: (*a*) peritectic at 1137 K at which $^L(4\cdot3\mathrm{Be}) + ^\alpha(2\cdot7\mathrm{Be}) \rightleftharpoons {}^\beta(4\cdot2\mathrm{Be})$; (*b*) eutectoid at 848 K at which $^\beta(6\cdot1\mathrm{Be}) \rightleftharpoons {}^\alpha(1\cdot4\mathrm{Be}) + ^\gamma(11\cdot3\mathrm{Be})$.

The α phase is a solid solution but the γ phase is an intermetallic compound based on CuBe.

The importance of this system lies in the rapid change in solid solubility of Be in Cu as temperature changes. This varies from

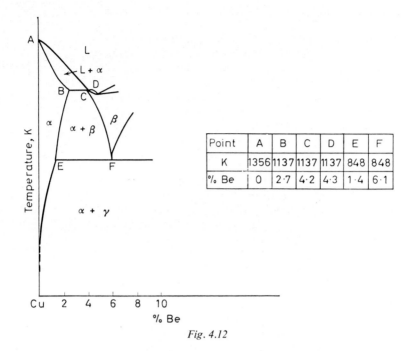

Point	A	B	C	D	E	F
K	1356	1137	1137	1137	848	848
% Be	0	2·7	4·2	4·3	1·4	6·1

Fig. 4.12

2·7% at 1137 K to less than 0·5% at room temperature and so age hardening is possible by sub-microscopic rejection of γ from supersaturated α produced by rapid cooling from about 1070 K.

The commercial alloys contain about 2%Be and may be cold shaped in between solution treatment and ageing so that the strengthening effect of cold work can be superimposed on that due to ageing. The alloys are used for the production of springs, diaphragms and non-sparking tools. The alloys are useful as high-strength wear-resistant electrical conductors.

Magnesium and its alloys

The pure metal is attractive because of its very low density ($1·74 \times 10^3$ kg/m³) but it has poor mechanical properties.

The metal has a CPH crystal structure and so plastic deformation is restricted to the basal {0001} planes. The metal becomes considerably more plastic above about 470 K, since at these temperatures other slip planes become operative. The lack of ductility at room temperature does, however, confer very good machining properties.

The metal has rather a low melting point (923 K) and is chemically active. It can, for example, undergo spontaneous combustion at elevated temperatures if the surface area is high. The metal and its alloys can, however, be welded without danger of combustion, since the high conductivity and specific heat prevent build-up of heat in the bulk material.

The alloying tendencies of magnesium are rather restricted. It is a highly electropositive element and readily loses 2 electrons per atom to give magnesium ions. There is, therefore, a strong tendency for the metal to chemically combine with added solutes rather than physically alloy with them. Most magnesium alloys, therefore, include intermediate compounds in the equilibrium diagram, and eutectics between the compound and the magnesium-rich primary phase are often formed.

Magnesium alloys of importance

Most alloys contain aluminium and the Mg–Al equilibrium diagram is given in part in Fig. 4.13.

The eutectic is a mixture of $(\alpha + \delta)$ and use of the lever rule indicates that it will contain more of the brittle $Mg_{17}Al_{12}$ compound than of the soft α phase. Alloys containing a eutectic network are therefore likely to be brittle.

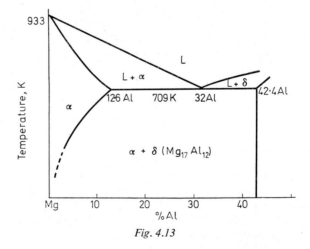

Fig. 4.13

The importance of the diagram is that the rapid drop in solid solubility exhibited indicates the possibility of age hardening.

Zinc is another important alloying element used with magnesium. Part of the Mg–Zn system is given in Fig. 4.14.

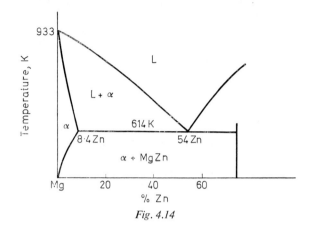

Fig. 4.14

Again this is a eutectic-type diagram with intermediate compound formation. Age hardening of the dilute alloys is a possibility.

Rare earth elements, i.e. the lanthanides from group IIIB of the Periodic Table, are also widely used as alloying elements. They are usually added in the form of *mischmetal*, containing about 50% cerium with 50% mixed rare earths, and confer improved properties at elevated temperatures. This improved creep resistance is probably due to the stiffening effect resulting from the precipitation of fine particles of Mg–rare earth compounds. Manganese is also used together with rare earth elements, since this retards the growth of the Mg–rare earth compound particles.

Casting alloys of magnesium are widely used. Since magnesium is highly reactive, the alloys need to be melted out of contact with air. Usually a chloride flux is floated on the metal as protection. The liquid alloys will also react with moisture in sand moulds giving oxidised and porous surfaces. Inhibitors such as sulphur or boric acid are added to moulding sands to prevent this.

Most of the alloys are based on Mg–Al–Zn and can be age hardened. The phase diagrams indicate, however, that the cast alloys are likely to be rather brittle because of the presence of eutectic. Solution treatment and ageing tends to remove the interdendritic network of eutectic found in the as-cast condition and so improves the ductility as well as increasing strength.

Wrought alloys are again based on Mg–Al–Zn and the alloys are again age hardenable. Because of the limitations on plastic deformation imposed by the CPH crystal structure, much of the mechanical shaping is done at temperatures above 470 K.

Solution treatment of magnesium alloys is carried out from about 690 K by air cooling. Ageing then proceeds at about 440 K. Since the alloys are, if necessary, hot worked at 570–620 K, air cooling after hot working gives something approaching a solution-treated condition and this is one of the reasons why worked alloys are more ductile than cast alloys.

Nickel and its alloys

Pure nickel is a FCC metal having good corrosion and oxidation resistance and retaining a good proportion of its strength at elevated temperatures. It is also a ferromagnetic material. Properties such as these make nickel, and particularly its alloys, attractive as engineering materials.

Nickel is rapidly attacked by sulphurous environments above about 530 K, and above 913 K a Ni+NiS eutectic is produced which may lead to intercrystalline embrittlement. Magnesium is usually added to desulphurise and prevent this effect.

The commercial grades of pure nickel are designated 'A' or 'L', depending on carbon content, as indicated in Table 4.4.

Table 4.4

Grade	C	Fe	Mn	Si	C	S
A	0·1	0·15	0·2	0·05	0·01	0·005
L	0·02	0·05	0·2	0·15	0·01	0·005

The low carbon grade L is soft and does not work harden readily. It is preferred for cold-working operations.

Most of the nickel produced is used in the manufacture of alloys, although commercially pure material is used in electroplating as an undercoating for chromium plating.

Nickel–Iron alloys are important because of their magnetic and thermal expansion properties. Many of these alloys have remarkably low coefficients of expansion and are used as metrological standards, temperature controllers, glass-metal seals, etc. Pure nickel, for example, has an expansion coefficient of about $12 \cdot 5 \times$ $\times 10^{-60}$ C^{-1}, while an alloy of 35Ni : 65Fe has a coefficient of only about 1×10^{-60} C^{-1}.

The main group of nickel–iron alloys of magnetic interest lies in the 35–80%Ni range. These alloys are magnetically soft, i.e. they are not permanent magnets but they have high permeability and are used in light transformers and armatures to replace iron.

The addition of aluminium and cobalt to Ni–Fe alloys produces a series of magnetically hard permanent magnet materials which give very high magnetic strength in small volumes. These are the Alnico and Alcomax alloys.

Nickel–Copper alloys have outstanding corrosion resistance and are widely used in chemical and food plant, pickling equipment and marine power plant. The straight Ni–Cu alloys are all single phase (see Fig. 4.10) and are marketed under the trade name Monel metal, some of which can be made age hardening by addition of aluminium to produce Ni_3Al intermediate compound.

Nickel–Chromium alloys are of the *inconel* type containing about 76%Ni. The alloys are resistant to corrosion and wear particularly at elevated temperatures in oxidising, carburising or nitriding atmospheres. Fig. 4.15 gives the phase diagram for this system.

The diagram indicates the alloys as being single-phase α and so they can be mechanically shaped. Casting alloys containing silicon to improve fluidity are also available and age hardenability may be induced by additions of aluminium and titanium to produce Ni_3AlTi intermediate compounds.

The *Nichrome* alloys, containing either 80Ni : 20Cr or 60Ni : 24Fe : 16Cr, are also available as resistance heating elements.

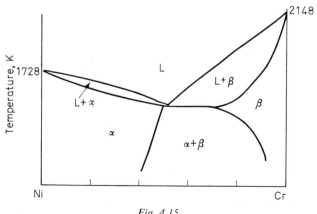

Fig. 4.15

Table 4.5

Alloy	Specification	C	Si	Cu	Fe	Mn	Cr	Ti	Al	Co	Mo	B	Zr	Pb	S	Ni
Nimonic 75	DTD 703B	0·081 0·15	1·0 max.	0·5 max.	5·0 max.	1·0 max.	18/21	0·2/0·6								bal
Nimonic 80A	DTD 736A	0·1 max.	1·0 max.	0·2 max.	3·0 max.	1·0 max.	18/21	1·8/2·7	1·0/1·8	2·0 max.		0·008 max.			0·015 max.	bal
Nimonic 90	DTD 747A DTD 5027	0·13 max.	1·5 max.	0·2 max.	3·0 max.	1·0 max.	18/21	1·8/3·0	0·8/2·0	15/21				0·005 max.		bal
Nimonic 93	None	0·1	1·5 max.	0·1 max.	2·0 max.	1·0 max.	19·5	2·75	1·5	18·0	0·3 max.	0·008	0·08			bal
Nimonic 115	DTD 5017	0·2 max.	1·0 max.	0·2 max.	1·0 max.	1·0 max.	14/16	3·5/4·5	4·5/5·5	13·5/ 16·5	3·0/5·0			0·005 max.		bal
Nimonic PE16	DTD 5047	0·1 max.	0·3 max.		bal	0·2 max.	15/18	0·9/1·5	0·9/1·5	2·0 max.	2·5/4·0 max.	0·005 max.	0·05 max.			42/45

The Nimonic alloys are rather complex alloys based on nickel, iron and chromium and have been designed specifically for use as creep-resistant materials. The basis of a good creep-resistant alloy is a relatively complex solid solution containing a number of different metals having the widestpossible dissimilarity in atomic sizes. The constituent metal should also have high melting points.

Such a combination produces a high-strength material in which dislocation motion is severely curtailed because of lattice strains. The material will also maintain this strength at elevated temperatures because of the high recrystallisation temperature. Many of the Nimonic alloys can also be age hardened and so this also reduces creep at elevated temperatures. Loss of strength due to over-ageing at operating temperature is, of course, a possibility. However, the driving force for actual precipitation in an age-hardened alloy is the interfacial energy at the particle–parent phase interface. If the degree of misfit between the precipitate structure and the parent phase structure is only slight, then precipitation is not so likely to occur. In the Nimonic alloys, there is very little difference between parent phase and precipitate phase crystal structure and so over-ageing is not normally a problem (see Chapter 8). The intermediate compounds responsible for ageing in Nimonics are of the type Ni_3Ti, Ni_3Al, Ni_3TiAl, and carbides and nitrides of titanium and chromium.

Typical compositions are given in Table 4.5.

Where age hardening is carried out, solution treatment temperatures are high, usually over 1270 K. Likewise, ageing tem-

Table 4.6

Temperature K	0·1 % PS MN/m²	UTS MN/m²	El. %	
473	293·6	730	39·8	⎫
773	277·5	693	37·4	⎬ Nimonic 75
1073	103·5	200	85·1	
1273	41·7	74·1	100·5	⎭
473	746	1202·8	25·0	⎫
773	710·2	1111·8	23·6	⎬ Nimonic 93
1073	526	616·1	34·5	
1273	49·4	95·7	97	⎭
473	803	1165	25·0	⎫
773	787·4	1158	25·5	⎬ Nimonic 115
1073	716	1019	20·0	
1273	216	431·5	24·0	⎭

peratures are high and so the alloys may be operated under load at elevated temperatures. Table 4.6 indicates the properties of some of the alloys at various temperatures.

Lead- and tin-base materials

The important materials here are the white metal bearing alloys which consist basically of hard particles of intermediate compound embedded in a soft matrix. In use, the hard particles are usually quoted as supporting the rotating shaft or member while the softer matrix, being worn away at a higher rate than the hard particles, provides a network of oil channels. This view of the bearing action of duplex structures is not always valid. In some leaded copper bearings, for example, the matrix is harder than the precipitated particles.

Typical bearing alloys are:

High tin alloys

These are based on the system tin–antimony–copper and are substantially free from lead. They contain 1–10% of copper with up to 18% antimony and have a microstructure consisting of a back-

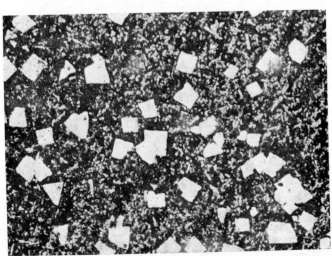

Fig. 4.16. Tin-base bearing metal—SnSb cuboids in a background of eutectic composed of a tin-rich solid solution and Cu_6Sn_5 ($\times$ 250)

ground of soft solid solution of antimony in tin, in which are embedded hard particles of intermediate compounds such as Cu_6Sn_5 and SbSn.

The SbSn particles are less dense than the liquid alloy and so gravity segregation may occur on slow freezing of the alloys. The entangling network of Cu_6Sn_5 particles which may also be present will prevent this segregation to some extent, but chill casting is usually needed.

The mechanical properties of these alloys, like all lead and tin alloys, are rather poor and some form of backing is needed to support the bearing shell.

Fig. 4.16 illustrates the microstructure of a bearing alloy.

High lead alloys

These are based on the lead–antimony–tin system and have the usual duplex microstructure of hard particles in a softer background. These lead-rich bearings are basically cheap substitutes for tin-base materials. They soften more readily at elevated temperatures, have lower mechanical strength and lower thermal conductivity than the high tin bearings. In general, they contain 12–18% antimony with up to 12% tin.

Tin-rich alloys were the first to be used for die casting because of their low melting point and high fluidity. They are, however, more expensive than, for example, zinc-base die-casting alloys and so are only used now if they can offer special properties such as better corrosion resistance or better dimensional accuracy. Typical compositions are given in Table 4.7.

Table 4.7

Compositions				Uses
Sn	Sb	Cu	Pb	
92	5·5	2·5		Pewter
83·3	8·3	8·3		Instrument cases
75	20	5		Small gears for accounting machines, etc.
75	10	3	12	General purpose die-casting alloy

Fusible alloys based on tin may have solidus temperatures as low as 340 K. They are alloys of tin with bismuth, lead, cadmium and zinc and are usually of eutectic composition. The alloys are used

9

in temperature safety devices, as low temperature solders, for foundry patterns and, because some of them which contain Bi expand on freezing, as packing and sealing rings.

Solders form one of the largest uses of tin. It is a characteristic property of tin that it will wet and adhere to the surfaces of metals like iron, copper and their alloys at relatively low temperatures. Solders usually contain lead as well as tin, but lead has no ability to wet the surfaces to be joined. It does, however, serve a useful purpose since its addition reduces cost and also introduces a paste range, i.e. produces solders with variable melting ranges, since the Pb–Sn system is a eutectic system. Antimony is also a common addition to lead–tin solders to give extra strength. Some typical properties of solders are given in Table 4.8. The balance in all cases is lead.

Table 4.8

Tin	Antimony	Solidus K	Liquidus K	Sp.gr. kg/m³	Uses
95	0	456	496	$7 \cdot 5 \times 10^3$	Joints in electrical instruments
65	0	456	459	$8 \cdot 4 \times 10^3$	High-strength solder
63	0	456	456	$8 \cdot 4 \times 10^3$	Fully eutectic. High-fluidity solder
60	0	456	462	$8 \cdot 5 \times 10^3$	Tinsmiths' work
50	0	456	487	$8 \cdot 9 \times 10^3$	General purpose
45	2·3–2·7	458	488	$8 \cdot 9 \times 10^3$	Antimonial solder
30	1·0–1·7	458	521	$9 \cdot 6 \times 10^3$	Plumbers' solder

Zinc and its alloys

The two most important uses for zinc are for galvanising and for the production of die castings. Zinc–base die castings are very widely used, particularly in the automobile and domestic machinery fields, but for satisfactory performance the zinc used must be of very high purity otherwise the castings are prone to intercrystalline corrosion. Typical compositions of commercial zinc are given in Table 4.9.

The metal itself is of CPH crystal structure with a wider than usual spacing between the basal planes. As a consequence of this elongated unit cell structure the metal exhibits marked anisotropy in physical properties. Cast zinc nearly always produces long columnar crystals with the basal planes aligned parallel to the long

Table 4.9

	Pb	Fe	Cd
High purity for die casting	0·006 max.	0·005 max.	0·004 max.
General grade	0·2 max.	0·03 max.	0·5 max.
Brass-making zinc	0·6 max.	0·03 max.	0·5 max.

axes of the crystals and this coarse structure may give difficulty in subsequent forming operations.

Aluminium is the chief alloying element in zinc-base die-casting alloys. There are two main alloy compositions covered by B.S.S. 1004 : 1955.

	Alloy A	*Alloy B*	
Aluminium	3·8 – 4·3	3·8 – 4·3	Alloying additions
Copper		0·75–1·25	
Magnesium	0·03–0·06	0·03–0·06	
Iron	0·10 max.	0·1 max.	
Copper	0·10 max.		
Lead	0·005 max.	0·005 max.	
Cadmium	0·005 max.	0·005 max.	Impurities
Tin	0·002 max.	0·002 max.	
Thallium	0·001 max.	0·001 max.	
Indium	0·0005 max.	0·0005 max.	

The zinc-rich end of the Zn–Al system is shown in Fig. 4.17. The diagram indicates that the alloys freeze via a $L+\beta$ paste range with solidification being completed by eutectic reaction. The α' in the eutectic is unstable and at 548 K decomposes by a eutectoid reaction to give $\alpha+\beta$. The final structure is thus primary $\alpha+$eutectoid of $\alpha+\beta$.

As such an alloy ages, structural changes occur. When freshly cast, the β contains about 0·35% Al but, on ageing, much of this Al is rejected in the form of the α phase so that the β eventually contains only about 0·07% Al. These precipitation changes occur at room temperature over a period of a few weeks and a slight shrinkage results. This dimensional change due to α precipitation can be allowed for and is usually speeded up by reheating to 423 K for a few hours after casting.

9*

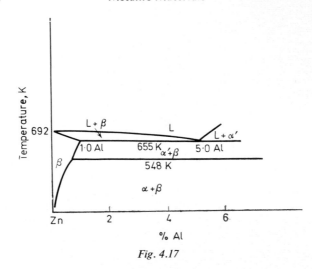

Fig. 4.17

The mechanical properties of zinc-base die-casting alloys depend largely on the way they have been cast. Pressure die castings with their inherent fine-grain structures tend to be better than sand or gravity die castings. The properties also depend on the temperature of testing since zinc and its alloys undergo a ductile–brittle transition as the temperature is reduced (see Chapter 8).

Even at sub-zero temperatures, however, the alloys are still sufficiently shock resistant for most purposes and, of course, the original toughness is regained as soon as the temperature rises again.

Aluminium and its alloys

Aluminium is a light metal (sp.gr. $2 \cdot 7 \times 10^3$ kg/m³) and weight for weight is a better electrical conductor than copper. The metal and its alloys are reasonably corrosion resistant because of the presence of a thin adherent surface coating of Al_2O_3. Such properties make the materials desirable engineering materials, particularly where weight is a penalty.

The pure metal is rather weak but by alloying and age-hardening treatments, strengths better than those of mild steel can be obtained so giving very favourable strength/weight ratios. These favourable ratios pertain only at normal temperatures, however, and the alloys cannot compete with steels under creep conditions.

There is a vast number of aluminium alloys available but these

Table 4.10

Specification	Cu	Mg	Si	Fe	Mn	Others	Condition	UTS MN/m²	El. %	Uses
NS 1	0·1		0·5	0·7	0·1		O	84·9	35	Panels and holloware
							H	154·4	5	
NS 4	0·15	1·75–2·75	0·6	0·75	0·5	Cr 0·5	O	185·3	24	Superstructures
							$\frac{3}{4}$H	255	4	
NS 5	0·15	3–4	0·6	0·75	1	Cr 0·5	O	216·0	18	Vehicle bodies
							$\frac{1}{4}$H	293·4	8	

can be broadly divided into heat-treatable and non-heat-treatable types. In the U.K., general engineering alloys are covered by B.S.S. 1470–1477 and 1490, aircraft materials by the B.S. 'L' series and by D.T.D. specifications.

Alloys for casting (B.S.S. 1490) usually contain silicon and are numbered 1 to 24 according to chemical composition. The number is prefixed by the letters L.M. and is followed by other letters to indicate the condition of the casting. The standard nomenclature is contained in the relevant specifications.

Non-heat-treatment wrought alloys

These alloys are used for the cold fabrication of panels, sheets, marine superstructures and the like. Because maximum ductility is needed, the alloys are usually dilute α solid solutions. Magnesium is the chief alloying element. The phase diagram for the Al–Mg

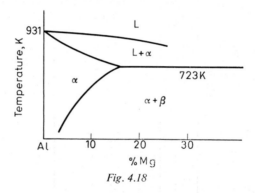

Fig. 4.18

system is given in Fig. 4.18 and, since commercial alloys contain less than about 5% Mg, the diagram indicates that such alloys will be mainly α solid solution. Table 4.10 gives the properties of some typical alloys.

Non-heat-treatable casting alloys

These are used both as sand and die castings and usually contain silicon around the eutectic composition to give maximum fluidity. The phase diagram of the Al–Si system is given in Fig. 4.19.

The lever rule indicates that the 10% Si alloy in the as-cast condition would consist of about 80% of eutectic and 20% ductile α.

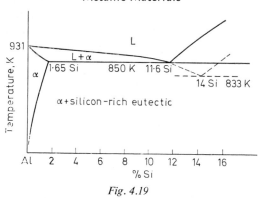

Fig. 4.19

The eutectic is rather coarse and such an alloy would give a tensile strength of 108–154 MN/m², with elongation = 3–5%.

By supercooling the alloy, the eutectic reaction is made to occur at lower temperatures and at higher silicon concentrations, as shown by the dotted lines. This depression of the eutectic produces a much finer eutectic and its proportion will now be only 10/14 ×100 = about 71%. The resultant properties would be, tensile strength 154–232 MN/m², with elongation 10–17%.

Such a modification of the eutectic could be produced by chill

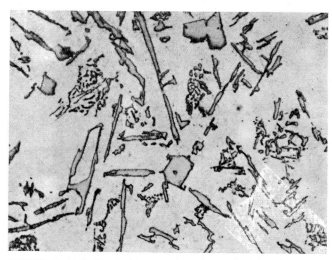

Fig. 4.20(a). Al–12% Si *alloy, unmodified—coarse needles of silicon-rich phase forming part of the eutectic* (× 250)

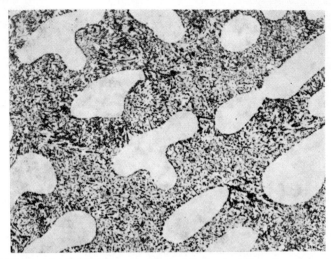

Fig. 4.20(b). Al–12% Si *alloy, modified—primary dendrites of* α *phase rich in* Al *with finely divided eutectic* (× 250)

casting but the undercooling effect is more easily produced by adding small quantities of sodium to the melt just before casting. The sodium probably fluxes out solid impurity particles which would normally act as seeds for crystallisation, and so freezing is delayed until lower than usual temperatures. The modifying effect is lost on subsequent remelting. Figs 4.20(a) and 4.20(b) show the unmodified and modified structures, respectively, while Table 4.11 gives the properties of some typical alloys.

Heat-treatable wrought alloys and casting alloys

Alloying elements such as copper, silicon, iron, nickel, etc., when added to aluminium, produce systems which are capable of being age hardened. The equilibrium diagrams are usually of the simple eutectic type and indicate second phases which are intermediate compounds. The phase diagram for the Al–Cu system illustrates this (Fig. 4.21).

This Al–Cu system forms the basis of the duralumin alloys, and copper contents close to the solubility limit would be chosen to give the maximum strengthening effect.

With the cast alloys, solution treatment is often dispensed with because the cooling rate during casting produces the same effects.

Table 4.11

Specification	Si	Cu	Mg	Mn	Zn	Condition	UTS MN/m²	El. %	Uses
LM1-M	3	7			3	Chill cast	169·5	1	Gravity die castings
LM2-M	10	1·6				Chill cast	247	2·5	Pressure die castings
LM5-M			4·5	0·5		Chill cast	200·7	10	Highly stressed castings
LM6-M	12					Chill cast	216	10	Automobile castings

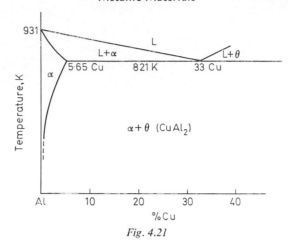

Fig. 4.21

These alloys are simply reheated after casting to give the strengthening effect.

Some of the intermediate compounds responsible for age hardening in aluminium alloys are $CuAl_2$, Mg_2Si, $FeAl_3$, $NiAl_3$.

Manganese and titanium are often added to prevent grain growth during solution treatment. Some typical alloys are detailed in Table 4.12.

Titanium Ti and its alloys

Titanium is one of the so-called newer metals and is coming into wider use because of its combination of low density and, in the alloyed form, high strength. The materials are expensive, this high cost being a reflection of the difficulties in extracting and fabricating. The oxide of titanium cannot be reduced by normal reducing agents because of its very high stability and so the oxide must first be converted into a chloride, this chloride being reduced by metallic magnesium in an argon atmosphere. Such a process yields crude solid titanium and further complex processes of purification must follow.

Fabrication presents further difficulties, since within the usual hot-working range, 800–1300 K, the metal rapidly picks up oxygen and nitrogen from the atmosphere and these can produce severe embrittlement. Fortunately, the diffusion rate of oxygen and nitrogen in titanium is low and so removal of the surface layers by pickling after hot working will prevent embrittlement.

Table 4.12

Specification	Cu	Si	Mg	Mn	Others	Treatment	UTS MN/m²	El. %	Uses
HE9-WP		0·5	0·6		0·2Ti	W 793 K / P 443 K/10h	247	18	Window frames
HE10-WP		1	0·8	0·5	0·5Cr	W 783 K / P 448 K/10h	247	20	Electric conductors
HF12-WP	2·3	1	1		1·0Ni 1·0Fe	W 803 K / P 443 K/15h	416·9	10	Forgings
HR13-T	2·2		0·3		0·3Ti	W 768 K / P 293 K/5 days	293·4	20	Rivets
HG15-WP	4·1	1	0·5	0·7	1·0Fe	W 783 K / P 443 K/10h	509·5	10	Aircraft parts
LM4-WP	3	5		0·5	0·8Fe 0·2Ti	W 793 K / P 443 K/12h	324·2	1	High-strength castings
LM13-WP	0·9	12	1·2		2·5Ni	W 793 K / P 443 K/10h	308·8	1	Pistons
LM14-WP	4	0·3	1·5		2·0Ni 0·2Ti	W 783 K / P 373 K/2h	277	0	Cylinder heads

Pure titanium

The metal is classed as a light metal, sp.gr. $4\cdot5\times10^3$ kg/m³. It is
interesting to compare the physical properties of the metal with
those of common constructional metals such as aluminium and
iron (Table 4.13).

Table 4.13

Property	Al	Ti	Fe
Density, kg/m³	$2\cdot77\times10^3$	$4\cdot5\times10^3$	$7\cdot87\times10^3$
Melting point, K	933	1941	1808
Thermal conductivity, W/m deg. C	238·66	17·17	72·85
Expansion coefficient, K⁻¹	$23\cdot5\times10^{-6}$	$8\cdot35\times10^{-6}$	$12\cdot3\times10^{-6}$
Modulus E, MN/m²	71×10^3	$106\cdot9\times10^3$	$196\cdot5\times10^3$

The high melting point of titanium promised usefulness as a
creep-resistant material but the reactivity of the metal at elevated
temperatures has prevented this application.

Pure titanium is very difficult to prepare because of the difficulty
in removing and preventing pick-up of impurities such as oxygen,
nitrogen, hydrogen, iron. Although it is a CPH metal, it is reasonably ductile when pure, but this ductility rapidly decreases in
commercially pure metal.

Both the metal and its alloys have very good corrosion resistance
as a result of the tight, impervious film of TiO_2 on the surface.

Titanium alloys

Titanium is an allotropic material, existing as CPH below 1155 K
and BCC above this. The importance of alloying elements lies in
their influence on this transformation.

The α-phase alloys (CPH) are produced by additions of aluminium and tin. Elements such as copper would also stabilise the
α phase but would, of course, detract from the high strength/weight
ratio inherent with titanium. Aluminium is lighter than titanium
and so, besides giving increased strength, Ti–Al alloys are even less
dense than pure titanium. The hardening effect of aluminium is so
pronounced that it must be limited to below about 7%. Addition

of 6% Al will raise the UTS from 310 to 590 MN/m² while reducing elongation from 40 to 20%. A small percentage of tin, e.g. Ti + 5 Al + 2·5 Sn gives further increase in strength without further reduction in ductility.

β-phase alloys (BCC) may also be produced by adding elements such as vanadium and molybdenum, but such large quantities of alloying element are needed that the strength/weight ratio is drastically reduced.

The most widely used titanium alloys are the $\alpha + \beta$ types, produced by adding small quantities of transition elements such as chromium, iron and nickel. These elements produce eutectoid systems as illustrated in Fig. 4.22.

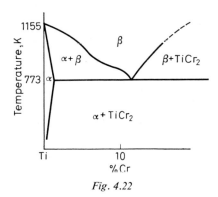

Fig. 4.22

The eutectoid change is very sluggish and does not normally occur. The alloys are usually hot worked in the β range and then annealed in the $\alpha + \beta$ range to control the proportion of $\alpha + \beta$ (use of lever law) followed by air cooling. Typical alloys are:

Ti–7% Mn,   Ti–3% Cr–1·5% Fe

Ti–2·75% Cr–5% Al–1·25% Fe

The ternary system Ti–V–Al, although not a eutectoid system, also includes an $\alpha + \beta$ region and a widely used alloy is Ti–6% Al–4% V giving 1300 Mn/m² UTS with 14% elongation.

Titanium alloys can give strengths equal to those of heat-treated alloy steel, have better fatigue resistance than mild steel, retain strength at elevated temperatures better than either aluminium alloys or mild steel, have outstanding corrosion resistance and are of low density. Because of these properties, titanium alloys are being used in aircraft, particularly military aircraft, to replace steel and aluminium. Replacing steel by titanium alloys could, for instance, convert a 60-seater aircraft into an 80-seater.

The metal is more expensive than stainless steel but, since it is possible to spot weld thin titanium sheet on to a cheap steel casing, it is possible to produce chemical vessels as cheap as those made entirely from stainless steel and having better corrosion resistance.

FERROUS MATERIALS

A ferrous material is one which contains a majority of the allotropic element iron. They form the vast majority of engineering metallic materials since they are both comparatively cheap and have high elastic moduli.

The allotropic behaviour of iron has been outlined in Chapter 1, and it was pointed out then that it was because of this allotropic behaviour that ferrous materials could be heat treated to vary the mechanical properties. This attainment of varying properties can now be discussed in more detail.

Steel and its heat treatment

Steel is basically an alloy of iron and carbon with carbon contents up to 1·7%. It may also contain alloying elements such as nickel, chromium, vanadium, and always contains residual impurities such as silicon, sulphur, manganese and phosphorus. Carbon is, however, the master element.

The heat treatment of steel consists of heating and cooling the solid at controlled rates in order to vary the microstructure and hence the mechanical properties. This is possible because of the allotropic behaviour of pure iron and of alloys based on iron. The allotropic changes of interest are those which occur below about 1270 K and so involve the changes from FCC austenite to BCC ferrite and vice versa. The changes are diffusion controlled and so need time to occur. They will thus only occur at the theoretical temperatures with very low rates of heating and cooling.

The effect of carbon additions on the allotropic behaviour of pure iron can be illustrated by the derived cooling curves shown in Fig. 4.23. The curves indicate that, as carbon is added,

(a) The upper arrest temperature is lowered below 1183 K. This upper arrest temperature is usually designated the A_3 temperature.

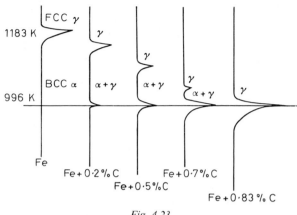

Fig. 4.23

(*b*) A new arrest temperature appears at 996 K (the A_1 temperature), which is constant regardless of carbon content.

(*c*) The length of an arrest is proportional to the amount of material involved in the change. The A_3 arrest becomes progressively shorter as carbon content increases, indicating that less and less free ferrite is being produced from the austenite. Hence, more and more of the austenite must be transforming into the material associated with the A_1 arrest. This A_1 arrest is obviously a eutectoid transformation.

With commercial rates of heating, there is a tendency to overshoot the A temperatures by about 30 K and so the arrest temperatures would be referred to as Ac_1 and Ac_3. Similarly, at normal cooling rates, the arrests occur about 30 K lower than usual and are referred to as A_{R_1} and A_{R_3}.

Carbon exists in steel in a number of different forms:

(*a*) A tiny amount of carbon exists in interstitial solid solution in the BCC ferrite lattice. This has been quoted as being less than $10^{-7}\%$ at 293 K.

(*b*) The vast majority of the carbon exists as an insoluble precipitate of the chemical compound iron carbide (Fe_3C or cementite). This intermediate compound contains 6.67% C, 93.33% Fe and is glass hard and brittle. 6.67% is the maximum amount of carbon that iron can absorb and if this much were added to liquid iron, the resultant solid would be all cementite. Hence the relevant phase diagram for steel is not the Fe–C system but

rather the Fe–Fe₃C system. In this system, the terms ferrite
and austenite are still used but they now refer to solid solutions.
Thus ferrite or α may refer to pure BCC iron but it also refers
to any solid solution of, say, C, Si or Mn, etc., in BCC iron.

That part of the Fe–Fe₃C system of interest in the heat treatment
of steel is given in Fig. 4.24 which is constructed from the derived
cooling curves given previously.

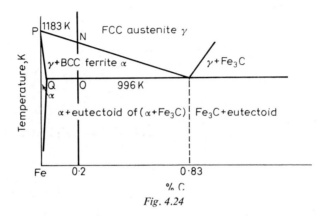

Fig. 4.24

The eutectoid structure deposited from the austenite on cooling
to 996 K is a laminated mixture of ferrite and cementite and is
called pearlite. The pearlite contains 0·83% C, as did the austenite
from which it was produced.

Consider the slow cooling of mild steel (0·2% C) from above its
A_3 temperature N. Between its A_3 at N and its A_1 at O, the γ
progressively changes to α. Using the lever law, it appears that the
composition of the α will vary down the line PQ. Composition Q
lies at about 0·03% C and so it is evident that when an austenite
grain containing 0·2% C in solid solution changes to ferrite, it
must reject most of its carbon. This rejected carbon is at once
dissolved by the austenite grains which have not yet changed to
ferrite and so, as cooling goes on, the remaining austenite becomes
enriched in carbon. At A_1, the austenite which is left in the structure
will contain 0·83% carbon and becomes saturated. Any further
attempt to cool this enriched austenite causes it to transform to
ferrite and reject carbon as Fe₃C. This gives simultaneous precipitation of α + Fe₃C in a lamella form and this is the eutectoid
pearlite. At room temperature, such a steel would contain
0·2/0·83 × 100 = 25% of pearlite and 75% of ferrite.

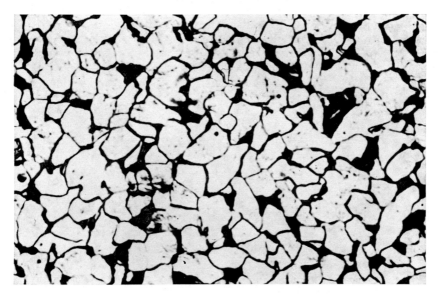

Fig. 4.25(a). 0·1% carbon steel, normalised

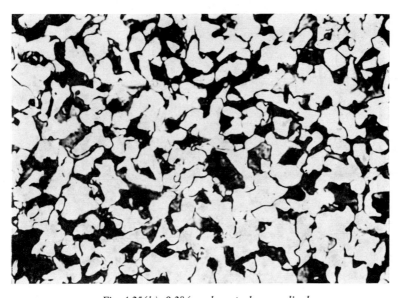

Fig. 4.25(b). 0·3% carbon steel, normalised

Fig. 4.25(c). 0·7% carbon steel, normalised

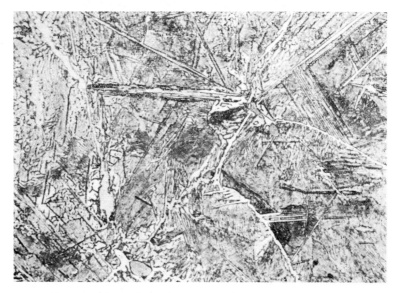

Fig. 4.25(d). High carbon steel, 1·0% C—grains of pearlite with free Fe_3C on the grain boundaries ($\times$ 250)

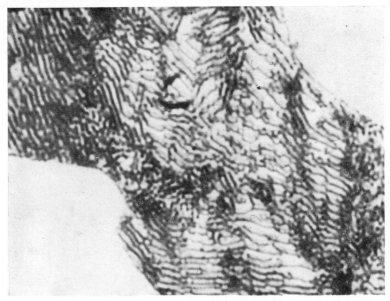

Fig. 4.25(e). Structure of pearlite in steel

As carbon content increases, the volume fraction of pearlite in the final structure will also increase, until at 0·83% C, the steel would have a fully pearlitic structure. Above this carbon content free cementite appears in the microstructure, i.e. Fe_3C, which is not associated with the pearlite laminations.

Figs 4.25(a) to 4.25(e) illustrate the influence of carbon content on microstructure.

Steel containing less than 0·83% C is referred to as hypo-eutectoid steel, while above 0·83% C the steel is know as hyper-eutectoid.

A straight iron–carbon alloy therefore contains two phases, α and Fe_3C, at room temperature. These phases have vastly different

Table 4.14

	Ferrite	*Cementite* Fe_3C	*Pearlite*
UTS, MN/m^2	293·4	54	895·5
El, %	50	0	10
VPN, MN/m^2	784·6	6374·5	2451·7

mechanical properties, and pearlite, which is a mixture of the two phases, will also have a separate set of properties. The figures in Table 4.14 illustrate this.

The presence of free cementite is therefore to be avoided. It is also obvious that the overall mechanical properties of an iron–carbon alloy will be a function of the ratio of ferrite to pearlite in the microstructure. This is illustrated in Fig. 4.26.

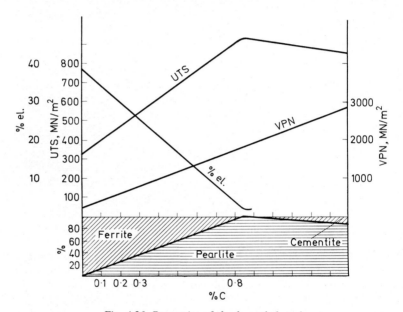

Fig. 4.26. *Properties of slowly cooled steel*

Influence of impurities in steel

In any steel there are minor amounts of impurities residual from the steelmaking process. Such elements may modify the properties if their quantity increases above normal values. These residual elements are:

Silicon. This is usually held below about 0·4%. It is present in solid solution in the ferrite and exerts a stiffening effect. Larger quantities may cause embrittlement.

Phosphorus. Is limited to below 0·05%, at which level it is in solution in ferrite. Large amounts would produce embrittling precipitates of Fe_3P.

Sulphur. Is limited to below 0·05%. It will exist in steel as a precipitate of either MnS or FeS. The presence of iron sulphide FeS is to be avoided since it freezes after the main body of the steel has solidified and so produces an embrittling grain boundary film. The presence of MnS is not so objectionable since this freezes at higher temperatures and does not take up grain boundary positions. Fortunately, MnS will form in preference to FeS provided the sulphur content is low enough. Sulphur may be deliberately added up to about 0·2% to give free machining steel.

Manganese. Will exist either as MnS or associated with the pearlite as Mn_3C. Its presence is not, therefore, undesirable.

Before the properties of steel can be altered by heat treatment, it is necessary to first obtain the steel in the austenitic condition. A necessary preliminary, therefore, is heating to Ac_3 or above. The heat treatment given may be designed to homogenise the material or to harden it. These treatments are now discussed.

Annealing

Full annealing is often applied to steel castings to homogenise the structure, remove the effects of segregation and remove casting stresses. An improvement in ductility and toughness results.

Cast hypo-eutectoid steel is usually characterised by a coarse, uneven structure called a Widmanstatten structure. This sort of structure is produced by cooling fairly quickly through a solid-state phase change such as the $\gamma \rightarrow \alpha$ change. The α tends to precipitate on to definite crystallographic planes and is not given time to diffuse. Such a structure often has low ductility. Fig. 4.27 illustrates this structure in cast mild steel.

Annealing such a steel simply consists in slowly heating the steel to Ac_3, soaking at this temperature to allow the carbon content to even out and then cooling with the furnace.

On reheating the cast steel, the pearlite in the steel transforms to austenite at Ac_1. This transformation nucleates at a number of points within each pearlite grain and so a single grain of pearlite will transform to produce many grains of austenite.

Between Ac_1 and Ac_3, the ferrite in the structure also transforms to austenite by the same process of nucleation. Thus, at Ac_3 the original coarse Widmanstatten structure will have been completely wiped out and replaced by fine equiaxed austenite. Soaking at Ac_3 is limited since grain growth of the austenite occurs rapidly and this is to be avoided since a coarse-grained austenite produced at Ac_3 tends to produce coarse-grained ferrite at room temperature.

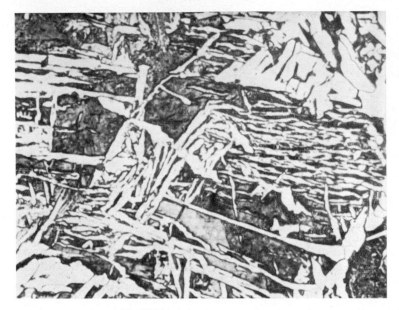

Fig. 4.27. Widmanstatten structure in cast steel

The soaking time is therefore limited to that needed to allow the high carbon content of the austenite produced from the pearlite to diffuse evenly throughout. Most of the grain refinement, therefore, occurs on heating. The cooling rate does have some influence on final grain size since more rapid cooling causes more nucleation centres of ferrite to be produced between A_{R_3} and A_{R_1}. This effect is, however, minor when compared with the refinement produced on heating.

On cooling, therefore, the equiaxed austenite changes back to equiaxed ferrite between A_{R_3} and A_{R_1} and finally to equiaxed pearlite at A_{R_1}. This replacement of the Widmanstatten structure by an equiaxed homogeneous structure produces considerable improvement in properties.

Normalising consists in heating the steel to Ac_3, soaking as usual and then air cooling. The changes in structure are those discussed under annealing but, because of the more rapid cooling from Ac_3, the final ferrite–pearlite structure is finer than in annealed steel. The normalised steel thus contains more grain boundary material and is as a result tougher and stronger. The pearlite laminations are also finer than usual and so crack propagation through the Fe_3C layer is less easy.

Influence of rapid cooling from Ac₃

The behaviour of austenite when it is cooled from Ac_3 at progressively increasing rates is illustrated in the derived cooling curves shown in Fig. 4.28 which refer to a 0·5% plain carbon steel.

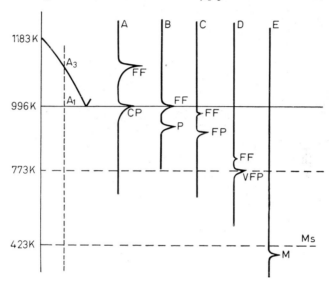

Fig. 4.28. Derived cooling curves of austenite when cooled from Ac_3 at progressively increasing rates (FF, free ferrite; CP, coarsely laminated pearlite; FP, finely laminated pearlite; VFP, very finely laminated pearlite; M, martensite)

The diagram indicates that at equilibrium cooling rates austenite will transform over the theoretical A_3–A_1 range with the production of free ferrite grains and coarsely laminated pearlite. Curves B, C, D, refer to progressively increasing cooling rates and, as expected, one result is the depression of the critical points to lower temperatures.

It is also evident from the lengths of the arrests, that the upper critical temperature is becoming less evident. This indicates that the amount of free ferrite is decreasing while the amount of pearlite is increasing. At some cooling rate just beyond D, the steel would have a fully pearlitic structure but the pearlite would be very finely laminated (VFP) and would not, of course, contain the usual 0·83%C. It could only contain 0·5%C. However, from a microstructural point of view, increasing the cooling rate is roughly equivalent to increasing the carbon content and this has the expected effect as regards resultant mechanical properties.

All transformations up to about cooling rate D involve a diffusion-controlled change

$$FCC\gamma \to BCC\alpha + Fe_3C \quad \text{precipitate}$$

However, if the cooling rate is so high that the austenite is trapped to below a temperature called the *Ms* temperature, it no longer transforms in this manner. The transformation occurs instantaneously by shearing of the FCC lattice to produce a body-centred tetragonal lattice and no precipitation of carbide occurs. This transformation will only go to completion if the temperature is continuously lowered. Thus, below *Ms*, austenite transforms by a diffusionless mechanism $FCC\gamma \to BCT$ and the carbon is still retained in solid solution. This BCT form is called *martensite* and is, of course, highly supersaturated with carbon. Dislocation motion within this supersaturated lattice is hardly possible and, as a result, the material is extremely hard and brittle. Its hardness increases as the carbon content of the steel increases.

Martensite is usually too hard and brittle to be of much use in engineering components but by reheating or tempering the quenched steel, a completely new microstructure may be obtained which is extremely tough. It is this structure which forms the basis of high-strength components and herein lies the importance of mar-

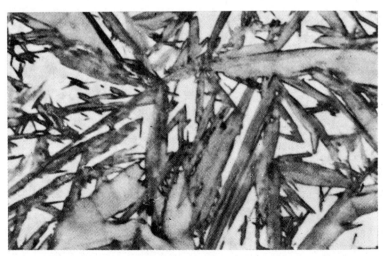

Fig. 4.29. Martensite plates in a high carbon alloy steel ($\times$ 2500)

tensite. In order to produce martensite in a steel, it is necessary to retain austenite to below the *Ms* temperature (which varies with the composition of the steel). Such retention of austenite in low carbon plain carbon steels usually involves very rapid cooling from the *Ac₃* and this, of course, introduces the risk of cracking and distortion due to differential thermal contraction. Such a steel is said to have a low hardenability. Addition of alloying elements to steel increases the hardenability, i.e. it becomes possible to produce martensitic structures at comparatively low cooling rates. Such additions may increase the amount of martensite produced but do not increase the hardness of the martensite. This is a function of the amount of carbon forcibly trapped in solid solution in the BCT lattice.

Under the microscope, martensite presents a characteristic acicular or needle-like appearance and is typically yellow in colour after etching. Fig. 4.29 illustrates this structure.

Tempering of martensite

The normal transformation of austenite produces BCCα+carbide precipitate. This is the stable condition. The abnormal transformation of austenite below *Ms* produces BCT martensite with no precipitation of carbide. One could thus regard BCT martensite as an intermediate lattice between FCCγ and BCCα. On this assumption, it is logical to expect that, by reheating the martensite, it would tend to revert towards the stable condition of BCCα+carbide precipitate. This reversion occurs during tempering in a number of stages:

(a) Between 373 and 473 K residual quenching stresses are relieved and rejection of carbon from the BCT lattice begins. The carbon is rejected in the form of a carbide of composition between Fe_3C and Fe_2C which reverts to Fe_3C at higher temperatures.

(b) Between 473 and 673 K more carbide is rejected and, as the supersaturation of the BCT lattice is relieved, it begins to revert back to BCC.

(c) At higher temperatures, reversion of BCT to BCC is completed and the carbide particles coalesce into larger globules.

The tempering temperature must, of course, be below A_1 otherwise austenite would be reproduced and this would give pearlite on subsequent cooling.

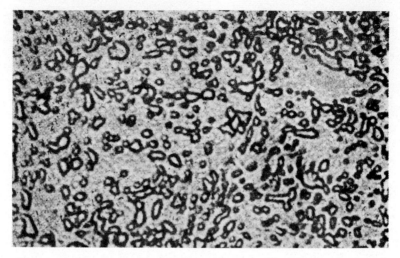

Fig. 4.30. Coarse sorbite—globules of carbide in a background of ferrite (×2500)

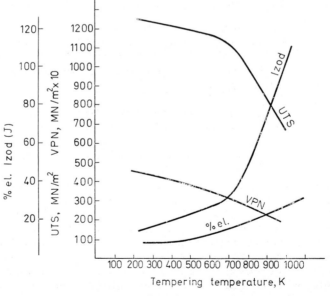

Fig. 4.31

The final structure and properties of a tempered steel will therefore depend on the tempering temperature. By tempering at low temperatures, some carbide rejection occurs but the acicular structure is still present. This is referred to as the tempered martensite structure and is hard, strong, but not very ductile. At higher temperatures the structure consists of a very fine distribution of carbide particles in a ferrite matrix. This is sometimes called the *sorbitic* structure and has very high toughness. Pronounced coalescence of carbide particles produces a spherodised structure and this over-tempered condition is usually avoided since it produces loss of strength. Fig. 4.30 illustrates a typical tempered structure.

The actual tempering temperature used will depend on the composition of the steel and the final properties required. As would be expected, steels containing alloying elements temper more slowly than plain carbon steels, since the presence of alloying atoms reduces diffusion rate.

Fig. 4.31 indicates the effects of tempering of martensite in a typical low alloy steel. The marked increase in toughness (Izod value) indicates why such microstructures are desirable in engineering components which have to withstand shock loading and high stresses.

The concept of hardenability and the influence of alloying elements in steel

It appears that tempered structures are desirable for high-strength components and it is obvious that such structures should be present throughout the whole body of the component. This means that the component must be capable of being made martensitic throughout, initially.

Diffusion rates in low carbon plain carbon steels are rather high and so to prevent austenite transformations above *Ms* requires drastic quenching with the attendant risk of cracking. Even with such high cooling rates, deep hardening of such steels is not possible and, after quenching, it is often the case that only the skin is martensitic, the centre being composed of ferrite + pearlite structures. Increasing the carbon content reduces diffusion rates and so increases depth of martensite formation, but high carbon steels are inherently non-ductile in any condition. By adding small amounts of alloying elements to low carbon steel, hardenability (depth of martensite formation) can be increased while still retaining the inherent ductility of low carbon contents. These alloying elements, such as nickel, chromium, molybdenum, etc., total about 5% in the general range of heat-treatable engineering steels and such steels can be cooled at comparatively low rates from Ac_3 yet still give fully martensitic conditions. In tool steels, the alloy content is considerably higher than this. The risks of distortion and cracking are thereby reduced and it becomes possible to 'through-harden' larger masses. The alloying elements, therefore, act by making the austenite transformation so sluggish that it becomes easy to retain the austenite to below *Ms*. The same sluggishness is also apparent in the decomposition of the martensite and so higher than usual tempering temperatures are needed. This effect can, of course, be utilised in the production of creep-resistant steels or steels which retain high strength and hardness at elevated temperature, i.e. tool steels. It is therefore important to be able to measure the heat treatment characteristics of a steel, i.e. to be able to predict the response of the steel to heat treatments designed to produce martensite. This sort of information is available for all the general engineering steels in the form of *S* curve data, Jominy curve data and continuous cooling transformation curve data.

'S' curves or 'time–temperature–transformation' curves

Normally, when a steel is heat treated, the cooling is continuous and the austenite transforms over a falling range of temperatures. Much useful information, however, can be obtained by studying the behaviour of austenite when it is made to transform at a constant temperature and this isothermal transformation behaviour is summarised in the '*S*' curve.

The *S* curve for a given steel is built up by treating thin discs of the steel. The first disc would be heated to Ac_3 to austenitise and would then be rapidly transferred into a molten salt bath held at the sub-critical temperature of interest, say 870 K. The steel, being of small mass, would cool almost instantaneously from Ac_3 to 870 K. After holding at 870 K for a given time, the steel disc would be finally quenched in water. Subsequent microscopic examination would then reveal how much, if any, of the austenite had transformed during the holding period at 870 K. For instance, if 10% of the austenite had transformed, the final microstructure would exhibit 90% of martensite and 10% of ferrite/carbide structure. Variation of the holding time at 870 K using different samples would then give data on the holding time needed for austenite transformation to begin and the time needed for transformation to be completed. By repetition of the tests with other samples at other temperatures, these holding times can be established for all temperatures below A_3.

The final results are then plotted in the form shown in Fig. 4.32, in which the line *O–P–Q* indicates the beginning of austenite transformation at any temperature and *R–S* outlines the end of the transformation. Between *O* and *P*, the austenite first changes to give ferrite, while the boundary *T–P–Q* indicates that the austenite is changing to structures such as pearlite or bainite.

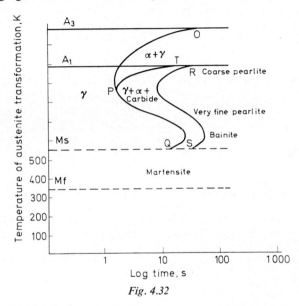

Fig. 4.32

Isothermal changes occur by nucleation and growth but it is nevertheless usual to superimpose on the diagrams the *Ms* and *Mf* temperatures. Martensite will only continue to be formed from austenite if the temperature is continuously falling. Thus *Mf* represents the finish of martensite formation.

The diagram clearly indicates the necessity for an incubation or waiting period at any temperature before transformation will begin. The extent of this waiting period depends on the composition of the steel and on the temperature of transformation.

One way of illustrating the use of *S* curves is to superimpose upon them continuous cooling rate curves. This has been done in Fig. 4.33. Curve *B* would represent a normalising cooling rate. The diagram indicates that the austenite begins to transform at 1 to give free ferrite. It continues to do this until 2 is reached, at which the remaining austenite produces pearlite by transforming between 2 and 3. Transformation is then complete and, since there

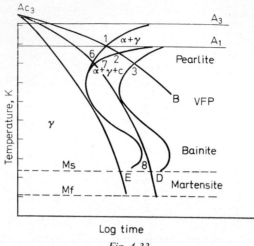

Fig. 4.33

is then no austenite left, the cooling rate thereafter cannot produce any further microstructural change. Curve *D* might represent oil quenching. A little free ferrite is produced between 6 and 7 while from 7 to 8 the remaining austenite transforms to give structures varying from pearlite to bainite. Since, at 8, the 100% transformation curve has not yet been crossed, there will still be some austenite left and this will then transform between *Ms* and *Mf* to give some martensite. This sort of mixed structure is typically found in large masses, i.e. martensitic skins varying to ferrite and pearlite in the centre.

The bainitic structures produced by transformations just above *Ms* are acicular in appearance rather like martensite. Unlike martensite, however, bainite is still based on a BCC structure and it is still produced by a diffusion reaction.

Curve *E* represents a drastic quench and it is obvious that the austenite is brought unchanged to *Ms–Mf* and so a fully martensitic condition should be produced. The cooling rate which just avoids cutting into the *S* curve is called the critical cooling velocity CCV and is the lowest cooling rate which will give a fully martensitic condition.

To obtain the best mechanical properties in a tempered steel, that steel should be fully martensitic before tempering. A knowledge of CCV is thus important.

It must be remembered that *S* curves are built up for isothermal conditions using very small masses of steel. In practice, austenite

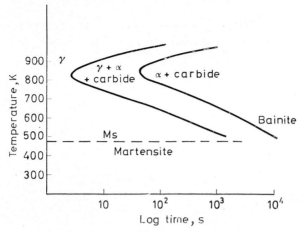

Fig. 4.34. S curve for En 42—0·75% *steel*

transformation is far from isothermal and large masses are involved. *S* curve data can therefore only serve as a guide in the selection of steel for any given application. The 'mass effect' is particularly important since the CCV for any given steel increases as the mass increases. In standard specifications (B.S.S.970) the mass effect is allowed for by use of the Ruling Section. This is the maximum mass which can be hardened and tempered to give the specified properties all the way through.

It is obviously beneficial when heat treating steel, to have a low CCV so that there is reduced risk of distortion and cracking during

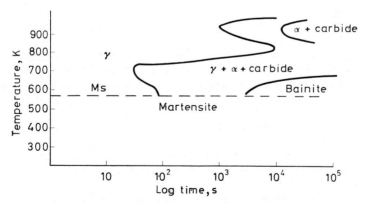

Fig. 4.35. S curve for En 24—0·4 C, 1·5 Ni, 1·20 Cr, 0·3 Mo

quenching. This reduction of CCV is one of the important influences of alloy additions to steel, since by reducing diffusion rates alloying elements will usually increase the incubation period. The increase may be such that the steel produces martensite simply by air cooling, i.e. air hardening steel.

This effect of alloying additions is illustrated in Figs 4.34, 4.35 and 4.36.

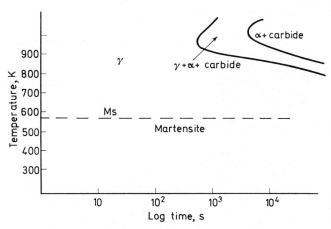

Fig. 4.36. S curve for En 56—0·24 C, 13·3 Cr

Continuous cooling transformation curves

These curves are based on *S* curve data but are more quantitative than *S* curves. A typical diagram is shown in Fig. 4.37, the horizontal axis of which refers to oil-quenching cooling rates. These curves are, however, available for air-cooling and water-quenching cooling rates also. The upper curve on the diagram represents the start of austenite transformation, while the lower curve marks the end of transformation.

To find the transformation behaviour at, say, mid-radius of a 4 unit dia. bar, a vertical line is drawn at this location. This then indicates that the austenite begins to transform at temperature *A* and is fully transformed at temperature *B*, the final structure being bainite+pearlite. Only bars less than diameter *X* would give fully martensitic structures. Again, the influence of alloying elements is to shift the curves bodily to the right.

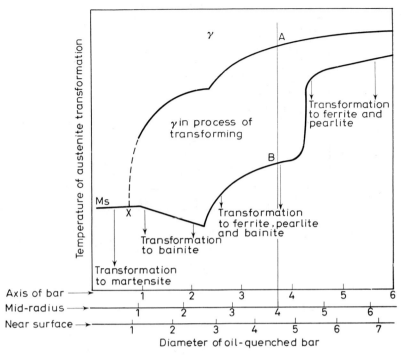

Fig. 4.37

Treatments based on S curves

If the *S* curve for a particular steel has the correct characteristics, it may be possible to apply treatments known as austempering and martempering.

In a normal direct quenching operation, the skin of the component cools more rapidly than the centre and it is this differential thermal contraction which leads to distortion and cracking.

In martempering, the austenitised component is quenched into a salt bath held at a temperature just above *Ms* and held there until the temperature differential is removed. The component is then finally quenched through *Ms–Mf*. Fully martensitic structures are still produced since no prior transformation of the austenite occurs above *Ms*. The treatment is sketched in Fig. 4.38.

In austempering, an interrupted quench is again used, but the component is held at the bainitic level and allowed to transform fully at this level. With some high carbon alloy steels particularly,

11

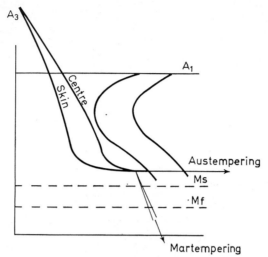

Fig. 4.38

a bainitic service structure is tougher and more creep resistant than would be a tempered martensite structure of comparable hardness. The austempering technique is also sketched in Fig. 4.38.

These treatments require accurate control and are only applied to small special components in which the extra cost is justified, e.g. die heat treatment.

Another treatment which can be applied to steel if it has a suitable S curve shape is that of ausforming. In this process, the steel is austenitised by heating above A_3 and then quenched to a temperature at which the S curve indicates a very long incubation

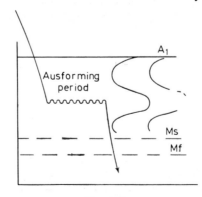

Fig. 4.39

period. While at this sub-critical temperature, the austenite is plastically deformed by rolling, forging, etc., and is then finally cooled through *Ms* to give martensite. The martensite produced from the strain-hardened austenite is very fine in structure and gives much higher strengths after tempering than conventional martensite, without a significant loss of ductility. Fig. 4.39 indicates the Ausforming process.

The plastic deformation of the austenite must be carried out in such a way and at such a temperature that processes such as recrystallisation have minimal effect. This condition, together with the requirement of a long incubation period on the *S* curve, places strict limitations on the types of steel which can be treated. The steels usually have compositions within the range:

$$C—0 \cdot 2–0 \cdot 5\%$$
$$Ni—1 \cdot 0–8 \cdot 0\%$$
$$Co—1 \cdot 0–4 \cdot 0\%$$

and will give, after ausforming and tempering, maximum tensile strengths up to about 3550 MN/m² with 6% elongation. In general, the lower the sub-critical temperature at which ausforming is carried out, the higher is the resultant strength. It is usually found that the actual strength level obtained is controlled by the carbon content, alloying elements playing only a secondary role.

Jominy hardenability data

Hardenability may be expressed quantitatively in terms of Ruling Section and could be defined as the depth to which martensite will be produced in a fixed mass under standard quenching conditions. Factors such as grain size, quenching technique, surface condition, type of coolant, etc., all influence hardenability and all these must be standardised so that the only variable is the steel composition. One widely used test which fulfils this condition is the Jominy test in which a 4 in × 1 in dia. bar is austenitised and then quenched from one end only under strictly controlled conditions. Within the same bar, therefore, are all cooling rates from drastic water quench at one end to air cooling at the other. These different cooling rates will be associated with different microstructures and mechanical properties and it is usual to take hardness readings along the end-quenched bar and plot these against distance.

Typical results are illustrated in Fig. 4.40. The curves show that plain carbon steels have much lower hardenability than alloy steels. Jominy curves for most steels are available and it is usual

11*

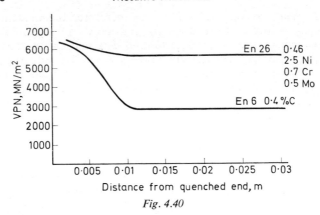

Fig. 4.40

to compare different curves, plotted on semi-transparent paper, in order that the most suitable steel may be selected for any proposed application.

High yield strength carbon steels

The yield strength is, or should be, the basis for engineering design and, since most of the steel used for large structures is mild steel, any method of raising the yield stress can result in useful savings in weight.

The yield stress of a mild steel may be raised from about 262 MN/m^2 to about 355 MN/m^2 by adding manganese up to 1·5%, but such steels (*a*) show a rapid drop of yield stress in thick sections on moving from outside to inside—this is simply a result of the excessive grain growth which occurs in heavy sections because of the high rolling temperatures needed and the slow cooling after rolling; (*b*) have rather low weldability and are not as tough as ordinary mild steel.

These deficiencies mainly stem from the fact that, being usually in the hot rolled condition, the ferrite grain size becomes increasingly coarse with increasing thickness of section.

Small additions of niobium (0·005–0·03%) can now be used to improve the yield stress of constructional steel (B.S.S.968: 1962). On cooling a niobium-bearing mild steel from hot rolling temperatures above 1470 K, carbides of niobium 'precipitate' but the 'precipitate' remains coherent with the ferrite lattice and so gives an age-hardening effect. This raises the yield stress but also, unfortunately, raises the ductile–brittle transition temperature (see Chapter 8).

By normalising the steel after rolling, the coherency of the precipitate is lost. However, the incoherent precipitate is fine enough to pin the austenite grain boundaries and inhibit grain growth. The resultant steel is therefore very much finer grained than usual and so still has a higher than usual yield stress, as would be indicated by the Petch relationship discussed in Chapter 2. The increased yield stress is not now associated with a raised ductile–brittle transition temperature. In fact, the transition temperature is lower than usual because of the finer ferrite grain size.

High alloy corrosion and scale-resistant steels

Corrosion and oxidation resistance is conferred by Cr_2O_3 which forms as an impervious self-healing skin over the surface of any material which contains chromium. Something over 11% Cr is needed in a steel before this skin is thick enough to be protective.

Apart from chromium, these high alloy resistant steels may contain other elements to confer special properties, but it is chromium which is the important constituent as regards corrosion resistance.

Chromium is a BCC element and so will tend to stabilise the ferrite phase. Chromium alone therefore could produce either a ferritic or a martensitic stainless steel depending on carbon content. Nickel is a FCC element and so will stabilise the austenite phase. Addition of enough nickel together with chromium could thus give an austenitic stainless steel.

Resistant steels, therefore, fall into two general types:

1. Hardenable steels

These are capable of producing martensite and so must be capable of undergoing the $\gamma \rightarrow \alpha$ transformation on cooling. Such hardenable steels are based on the Fe–Cr–C system and usually contain about 17% Cr with 0·25% C.

Fig. 4.41 gives part of the Fe–Cr system.

The adjusted diagram for 0·25% C indicates that the 17% Cr alloy should be ferritic below about 1173 K. In fact, hardenability is so high that martensitic structures are easily formed even on slow cooling (see S curves for En 56). Such a hard, stainless steel is obviously useful for knives, cutting tools, dies, etc.

By reducing the carbon content, the diagram indicates the possibility of producing ferritic stainless steel. Such a steel (17% Cr,

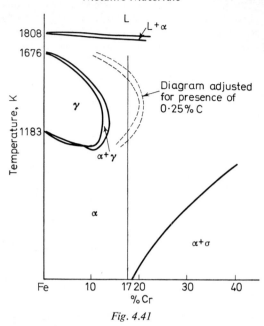

Fig. 4.41

0·1% C) cannot be quench hardened but, being completely ferritic, is ductile enough to be cold formed.

The σ phase indicated in the diagram is a brittle intermediate compound. The diagram indicates that it should not be present until chromium contents move above about 18–20% but, in practice, it can appear in alloys over 13% Cr if these are annealed for long periods. This sort of prolonged heating must be avoided in this type of steel in order to avoid embrittlement.

Like all stainless steels, the material has relatively high thermal expansion and low thermal conductivity. The higher carbon grade is basically an air hardening steel and so has very poor weldability. Being BCC, the material will also exhibit a ductile–brittle transition and so cannot be used for very low temperature applications (see Chapter 8).

2. Non-hardenable stainless steels

These are the austenitic steels, the austenitic condition being produced by addition of nickel. These steels cannot be made martensitic by quenching but severe cold work may cause the austenite to decompose to martensite and this can be a cause of embrittlement in the steel. Normal amounts of plastic deformation are acceptable, however, and in fact the material is often used for the production of sheet or strip which is then fabricated by welding.

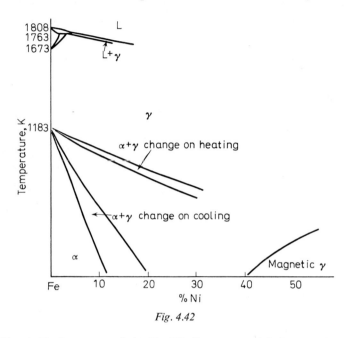

Fig. 4.42

Fig. 4.42 gives part of the Fe–Ni diagram and it is seen that the addition of Ni lowers the critical range, thereby stabilising the austenite phase. The depression of the critical range can be made more pronounced by increasing the cooling rate and in practice even a 7% Ni steel would be fully austenitic at room temperature at normal rates of cooling.

The effects of Cr in stabilising ferrite is opposed to the effect of Ni in stabilising austenite, and so by varying the Cr–Ni ratio it is possible to produce stainless steels of different microstructures. Fig. 4.43 illustrates this and refers to steel containing 0·1% C, rapidly cooled from 1370 K.

Austenitic stainless steel of the 18 Cr: 8 Ni type may cause trouble during welding if they have not been stabilised (see weld decay, page 308). Another difficulty with these steels may arise if they are heated for any length of time in the 770–1170 K range, since this induces precipitation of embrittling σ phase. This phase

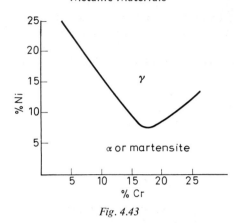

Fig. 4.43

precipitates from any ferrite which may be accidentally present in the steels and so one way of avoiding its presence is to use a composition well away from the γ/α boundary in Fig. 4.43. Thus a filler metal for welding austenitic stainless steel would have 35 Cr: 20 Ni composition and not 18/8.

Stainless steels of any type can be made precipitation hardening to give extra strength by the addition of elements such as copper and aluminium. In the martensitic types, very high strengths are obtainable by superimposing the strengthening effect of age hardening on to the strength of tempered martensite.

These age-hardened martensitic steels, the maraging steels, are, however, mainly based on the Ni–Co–Mo steels.

Maraging steels

Martensite can be produced in Fe–Ni alloys containing over 18% Ni, even with very low carbon contents, and such martensite is quite ductile, unlike Fe–C martensite. The martensite in these iron-nickel alloys is not supersaturated and contains a high density of mobile dislocations. It is produced from austenite without drastic cooling and so its formation involves no risk of distortion and cracking. The maraging steels develop their final very high yield strengths after age hardening of this ductile martensite.

Iron containing about 20% of nickel is austenitic at elevated temperatures and transforms to martensite on cooling at normal rates. There is, however, pronounced thermal hysteresis on the γ–martensite change. For example, martensite forms from the austenite on cooling to about 530 K but, on reheating, this martensite will only transform back to austenite at about 870 K. There is, therefore, an ample temperature range over which age hardening can be carried out and, in practice, an ageing temperature of about 750 K is used. The ageing tendency is induced by additions of elements such as molybdenum, which produces intermediate compound precipitation of Ni_3Mo.

Typical maraging steel contains 17–19% Ni, 7–9·5% Co, 3·0–5·0% Mo, 0·15–0·7% Ti with very low percentages of C, Si, S, Mn and P. After ageing, yield strengths of up to about 2162 MN/m^2 can be obtained.

The material, besides having ultra-high yield strength also has good toughness and corrosion resistance. In the martensitic condition before ageing, the steel is ductile enough to be cold worked (marforming). The rate of work hardening is low and cold work has very little effect on the hardness of the martensite. It does, however, increase the final strength after ageing and can lift the tensile strength by about 46 MN/m^2 for each 10% of prior cold reduction.

Since the martensite in these steels is produced from austenite without quenching, the steels have no hardenability in the accepted sense. Weldability is therefore good.

Maraging steels offer very high yield stresses and outstanding notch toughness, which is retained even at very low temperatures, and are finding use in applications where weight is a penalty, i.e. in aircraft and missile production. Because the high strength is produced by comparatively simple heat treatments involving no risk of distortion or cracking, they are also finding use as die materials and forming tools for metal working.

Tool steels

Steels for use in shearing, cutting and forming operations must have hardness, strength and wear resistance, together with adequate toughness. In many applications, heat resistance is also important.

Plain carbon tool steels contain 0·6–1·2% carbon, depending on the application. The lower carbon grades supply some toughness for use in chisels, etc., while higher carbon types are used for cutting edges. Plain carbon tool steels are comparatively cheap. The carbon is present either as martensite alone in the lower carbon grades, or as martensite with out-of-solution particles of carbide in the hyper-eutectoid types. Quenching is needed to develop hardness and there is some risk of distortion and cracking. Heat treatment atmospheres also need to be controlled to prevent surface decarburisation. The quenched steel is usually stress relieved at 420–600 K before use.

If high shock resistance or heat resistance is needed, alloy steels must be used. The function of the alloying addition is to increase hardenability, to produce wear-resistant out-of-solution carbides and to give resistance to softening at elevated temperatures. The alloying additions are therefore of strong carbide-forming elements such as chromium, tungsten, molybdenum, vanadium.

As alloy content increases, the necessity for rapid cooling to produce martensite decreases. High alloy tool steels are thus useful for die manufacture. With large quantities of elements such as tungsten or molybdenum, diffusion rates are so low that it becomes very difficult to temper the martensite. These are the high-speed tool steels which can operate at red heat without loss of hardness or cutting ability.

Cast irons

Cast iron can be defined as an alloy of iron and carbon, containing over about 2% carbon with varying amounts of Si, S, Mn, P. It may also contain deliberately added alloying elements.

The material is comparatively cheap and is used where strength and toughness are not of major importance. It is extremely 'castable' and has good machining properties and damping capacity.

The constitution of cast iron can usefully be studied using the iron–iron carbide equilibrium diagram, part of which has already been used in the discussion on steel. Fig 4.44 illustrates the relevant diagram.

Fe_3C is not a truly equilibrium condition since it can decompose into iron and graphite. The true equilibrium diagram is therefore the iron–graphite diagram. Fortunately, the metastable $Fe–Fe_3C$ system and the stable Fe–graphite system are very similar and both systems have been combined in Fig. 4.44. From this diagram it is obvious that, at room temperature, cast iron may consist of either (*a*) pearlite and cementite, (*b*) ferrite and graphite, or (*c*) pearlite and graphite.

The structure which exists in practice is a function of the chemical composition of the iron and the cooling rate during casting. If the carbon is present mainly as Fe_3C, then the iron will be extremely hard and brittle, i.e. white cast iron. Carbon present as graphite produces a soft, grey iron. This grey cast iron may still, however, be brittle if the graphite is present in flake form, because of the

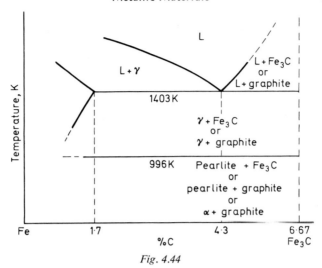

Fig. 4.44

stress-concentrating effect of the sharp-edged flakes which makes crack propagation easy.

Figs 4.45 and 4.46 illustrate the structures of white iron and flake graphite iron, respectively.

Alloys above the eutectic composition (4·3% carbon) are avoided in practice. If such a hyper-eutectic alloy freezes according to the Fe–Fe$_3$C system, then Fe$_3$C is the first phase precipitated from the liquid. If it freezes according to the Fe–graphite system, then graphite is the primary phase. In either case, the primary phase is freely floating in a liquid and has very little restriction on its growth. Very coarse structures are thereby produced and the resultant alloy is brittle and of low strength. Hypo-eutectic compositions are thus preferred.

Influence of chemical composition

Silicon in cast iron is a graphitising element. It may be present in amounts up to 3% and catalyses the breakdown of Fe$_3$C into iron and graphite: Fe$_3$C → 3Fe+C.

Increasing the silicon content of a cast iron can therefore convert it from a hard white iron to a soft grey iron. *Phosphorus* exists as the phosphide Fe$_3$P which is present as a binary Fe–Fe$_3$P eutectic in grey iron, or as a ternary Fe–Fe$_3$C–Fe$_3$P eutectic in white iron. Because of the presence of this eutectic, phosphoric irons may be

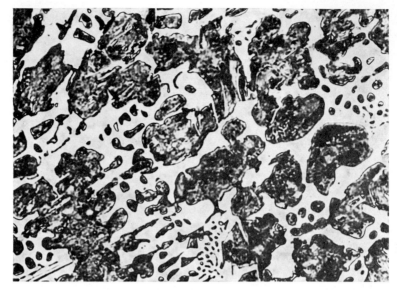

Fig. 4.45. White cast iron showing cells of pearlite in a background of cementite

Fig. 4.46. Graphite flakes in grey cast iron

used for production of thin or intricate castings. The eutectic, however, embrittles the material and so phosphorus is held at low levels in good-quality irons.

Both silicon and phosphorus reduce the carbon content of the eutectic below 4·3%. The extent of the shift in the eutectic position can be estimated using the carbon equivalent CE.

$CE = $ carbon $\% + \frac{1}{3}$ silicon $\% + \frac{1}{3}$ phosphorus $\%$. Thus an iron containing 3·4% C, 3·0% Si, 0·9% P would be hypo-eutectic if judged on carbon content only, but would, in fact, behave as if it contained $3\cdot4 + (3\cdot9/3) = 4\cdot7\%$ of carbon. It would therefore freeze as a hyper-eutectic iron and would be useless as an engineering material.

Manganese and sulphur may exist in cast iron as the sulphide MnS. If insufficient manganese or too much sulphur is present, the sulphur will produce iron sulphide FeS. This, as noted earlier, can cause embrittlement and so the sulphur content in good-quality iron is limited.

FeS is also a carbide stabiliser and prevents the decomposition of Fe_3C. A high sulphur content could therefore produce a white cast iron.

Influence of cooling rate

The cooling rate during casting is mainly a function of mass of casting and the nature of the mould material. Rapid cooling will retard the breakdown of cementite into iron and graphite and so a chilled, white iron is produced. This is evident in thin sections or in metal mould castings. Chilled iron is normally avoided because of its hardness and poor machining properties, but selected areas of a casting may be deliberately chilled to give wear resistance by the use of metal inserts in the mould.

General types of cast iron

The two chief types are white cast iron and flake graphite grey cast iron. Variations on these basic materials can be produced by the following treatments.

Inoculation

Good-quality iron of low sulphur and phosphorus content may be 'seeded' just before casting with solid particles of a compound such as calcium silicide. The increased density of nuclei for crystallisation induces the formation of much finer graphite flakes than usual, giving a higher strength iron.

Spherodisation

In this type of iron (nodular or spheroidal graphite), the normal flake graphite is made to crystallise in spheres by injecting the molten iron with magnesium and ferrosilicon just before casting. The magnesium is the active element, the ferrosilicon providing nuclei for crystallisation of the graphite. Removal of flake graphite removes stress concentration effects and the iron, as a result, is quite ductile. The as-cast structure consists of pearlite with nodular graphite. Annealing at about 1270 K decomposes the Fe_3C contained within the pearlite because of the silicon present and the final structure of ferrite with nodular graphite gives a further improvement in ductility. This type of iron is replacing cast medium carbon steel in many applications since it has rather similar properties. Figs 4.47(a), 4.47(b) illustrate the structures of these irons.

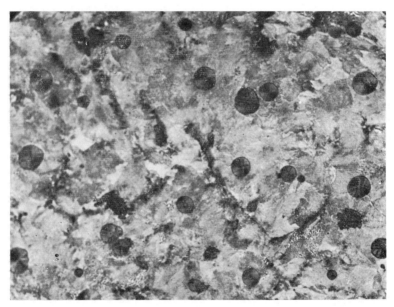

Fig. 4.47(a). Nodular cast iron in the as-cast condition—spheres of graphite in a background of pearlite ($\times 100$)

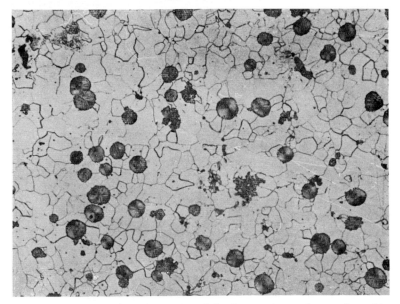

Fig. 4.47(b). Nodular cast iron annealed after casting—spheres of graphite in a background of ferrite (×100)

Malleableising

This is carried out by heat treatment of white cast iron. In the blackheart process, the white iron castings are annealed at 1120–1220 K for 50–200 h in a non-oxidising environment. In the presence of the contained silicon, the Fe_3C decomposes to give ferrite and small clusters of graphite. The iron, as a result, is quite ductile. In the whiteheart process, the white iron castings are annealed in an oxidising environment. The Fe_3C is decomposed as usual but the oxidising environment then burns out the clusters of graphite. In thin sections, therefore, whiteheart malleable iron can be fully ferritic but usually this is a skin effect only. Fig. 4.48 shows a typical structure of malleable iron.

Alloying

Alloying elements can be added to any type of cast iron. Nickel is widely used. It prevents grain coarsening in the heavier, more slowly cooled sections of castings and also has a mild graphitising

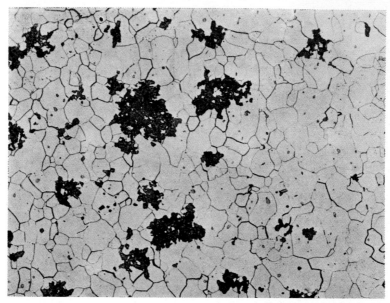

Fig. 4.48. Blackheart malleable cast iron—clusters of graphite in a background of ferrite (×200)

effect. Nickel contents of 2–5% introduce heat treatability and martensitic structures may be produced for wear resistance. With higher nickel contents of 10–30%, the iron remains austenitic on casting and is useful for corrosion resistance and non-magnetic applications.

Chromium is usually used in conjunction with nickel. It is a very powerful carbide-forming element and so induces chilling and white iron formation when present in large amounts. It may also confer wear and heat resistance.

Molybdenum is sometimes used in cast irons in conjunction with nickel, specifically to produce ferrite having an acicular or needle-like arrangement. These acicular irons are hard, fairly tough and wear resistant and are used for the manufacture of cast automobile crankshafts.

Typical properties of ferrous materials

Tables of mechanical properties of ferrous materials, although not given here, are readily available from British Standards Specifications or from the references listed in the Bibliography.

BIBLIOGRAPHY

AITCHISON, L., and PUMPHREY, W. I., *Engineering Steels*, Macdonald & Evans (1953)

Aluminium (Ed. Van Horn, K. R.), A.S.M. (1966)

Aluminium for Engineering Students, Aluminium Federation, London

BRICK, R. M., and PHILLIPS, A., *Structure and Properties of Alloys*, McGraw-Hill

BULLENS, D. K., *Steel and its Heat Treatment*, Battelle Memorial Institute, Wiley

Developments in Maraging Steels, International Nickel Co. Ltd, London

Engineering Materials (Ed. Sharp, H. J), Heywood, London (1966)

FORRESTER, P. G., *Babbitt Alloys for Plain Bearings*, Tin Research Institute (1963)

GEMMILL, M. G., *Technology and Properties of Ferrous Alloys for High-temperature Use*, Newnes, London (1966)

HOPKINS and RAY, 'Thermo-mechanical treatments', *Metal Treatment*, Vol. 30 (1963)

MCQUILLAN, A. D., and MCQUILLAN, M. K., *Titanium*, Butterworths, London

Mechanical Properties of Nickel Alloy Steels, International Nickel Co. Ltd, London

Mechanical Working of Steel (Ausforming), Met. Soc. Conferences, Vol. 21, Assoc. Inst. mech. Engrs (1963)

Metallurgical Developments in Carbon Steels, Iron and Steel Institute special report No. 81, London (1963)

MORROUGH, H., *Cast Iron as an Engineering Material*, Engineering Materials and Design Conference, London (1960)

PAYSON, P., *The Metallurgy of Tool Steels*, Wiley (1962)

PETTY, E. R., *Physical Metallurgy of Engineering Materials*, Allen and Unwin (1968)

Properties and Selection of Materials, A.S.M. Metals Handbook, Vol. 1 (1961)

Properties of Aluminium and its Alloys, Aluminium Federation, London

Properties of Tin Alloys, Tin Research Institute, London

REED-HILL, R. E. *Physical Metallurgy Principles*, Van Nostrand (1964)

ROLLASON, E. C., *Metallurgy for Engineers*, Arnold (1961)

SAMANS, C. H., *Metallic Materials in Engineering*, Macmillan, London (1963)

'The strength of steel', *Scientific American*, Vol. 209, No. 2 (1963)

Tool Steels, A.S.M. (1962)

Transformation Characteristics of Direct Hardening Nickel Alloy Steels, International Nickel Co. Ltd, London

Useful information on metallic materials is also obtainable from:

British Cast Iron Research Association
British Standard Specifications
Climax Molybdenum Co. Ltd
Copper Development Association
International Nickel Co. Ltd
Magnesium Elektron Ltd
Zinc Development Association

Chapter 5

POLYMERIC MATERIALS

As was indicated in Chapter 1, a polymer is composed of giant molecules produced by repetition of a simple unit called the mer. The polymerisation process may involve either addition or condensation reactions.

In addition polymerisation, the monomer has a double bond which is induced to break. The bond then liberated is free to add on to similar free bonds. Such a process can occur without any by-product and is the method by which polyethylene, polyvinylchloride, polytetrafluorethylene, etc., are produced. The addition reaction usually has to be initiated in some way, either by the presence of a catalyst or by application of heat and pressure.

If the polymer molecule is built up from one type of mer only, it is referred to as a homopolymer. Co-polymers are also possible, consisting of a number of different types of mer. This is analogous to alloying in metals and, just as an alloy may have better properties than a pure metal, so may more desirable properties be introduced by co-polymerisation. For example, polystyrene, a stiff, rigid material, can be co-polymerised with a rubbery compound called butadiene in order to produce a synthetic rubber of better rigidity than straight butadiene rubber, i.e.

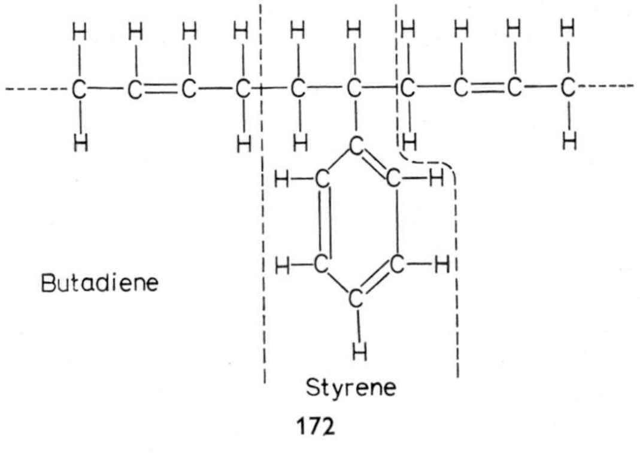

Butadiene

Styrene

In condensation polymerisation, a by-product is always rejected or condensed out from the reaction and the reaction does not depend upon the presence of multiple bonds.

The product of a condensation reaction need not be a linear molecule. For example, phenol formaldehyde, a bakelite-type plastic, is produced by a condensation reaction between phenol C_6H_5OH and formaldehyde HCHO, i.e.,

A network molecule is produced and one water molecule is rejected or condensed per reaction event. The condensation reaction may occur at a number of points around the benzene ring and so a tangled network is built up during polymerisation.

THERMOSETTING AND THERMOPLASTIC MATERIALS

In a network polymer such as phenol formaldehyde, there is the possibility of co-valent bonds being established between different parts of separate molecules. This 'cross-linking' effect occurs on heating and so such polymers, although they may be flexible when first produced, will set into a hard, rigid mass if they are heated. The action is not reversible and such polymers are widely used for the production of rigid mouldings.

A thermoplastic material such as nylon or polyethylene is one in which the molecules are linear. The only bonds between one chain and its neighbour are van der Waals' bonds and so, under stress, the chains can slide past each other, this flexibility becoming more evident as temperature increases. At some temperature, of course, the material will become fluid. All these changes are reversible. A thermoplastic would thus be expected to be generally tougher than a thermosetting polymer. Apart from classification based on

12*

thermoplastic or thermosetting behaviour, polymers may also be grouped according to the internal arrangement of their molecules. This gives rise to three main types which are discussed below.

1. Amorphous polymers (thermoplastics)

The long-chain molecules in this type are randomly interleaved and there is no crystallinity or cross-linking. One could visualise a mass of cooked spaghetti as an analogy. Such polymers tend to be glass-like and transparent and are typified by Perspex (polymethylmethacrylate).

Because of the random way in which the giant molecules have grown during polymerisation, chains of widely varying length will be present and so it is only possible to give an average molecular weight for the material. The molecular weight of a polymer controls such properties as viscosity and response to temperature changes, and it is evident therefore that in such a material it is possible to have a number of different physical states according to the temperature and the average molecular weight. Fig. 5.1 illustrates this.

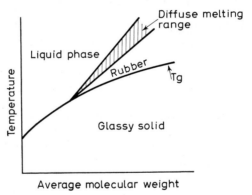

Fig. 5.1

At low molecular weights, a definite melting point exists but in the high molecular weight commercial polymers there is no such sharp transition between solid and liquid.

These polymers pass from a rigid to a rubber-like condition over a narrow range of temperature called the *glass transition temperature Tg*. This, of course, involves a complete and drastic change in mechanical properties.

At higher temperatures, the rubbery state transforms to liquid over a melting range.

The glass transition temperature Tg is often defined as that temperature at which, on cooling, molecular rotation about the C–C backbone bonds becomes restricted. Above Tg, in the rubber state, the molecules tend to take up a random coiled formation as a result of the tetrahedral distribution of the bonds on the carbon atoms, and the application of stress can then cause the uncoiling of the molecules as rotation about C–C bonds occurs. Removal of the stress allows the molecules to re-coil again but there is usually some hysteresis present, i.e. the deformation lags behind the application and removal of stress. Hence, the deformation characteristics of a polymer in its rubbery state above Tg will depend on the rate of application of stress. A material which behaves like a rubber when stress is slowly applied, may appear to be quite rigid and brittle if shock loaded. In effect, this corresponds to a raising of the apparent Tg temperature.

Stretching of an amorphous polymer while above Tg causes orientation of the molecules in the direction of stress and, if the polymer is allowed to cool below Tg with the stress still present, this orientation will be frozen in. The material then becomes anisotropic and has higher strengths parallel to the preferred orientation. Frozen-in orientations of this nature may result in distortion on subsequent reheating, since the molecules will re-coil as the stresses are released. This stress-induced orientation strengthening effect is widely used in the production of filament and film.

The glass transition temperature is an important concept when dealing with amorphous polymers since this temperature, in relation to the service temperature, will govern the mechanical behaviour of the material. Well below Tg, the material is rigid and capable of carrying stress. Above Tg, the material behaves as a rubber. The level of Tg is not constant but can be influenced by outside factors. Some of these factors are:

(*a*) The type of unit attached to the backbone. As a backbone bond rotates, different parts of the molecule move into different proximities to each other. The repulsive forces set up then influence the energy needed to cause rotation and hence the level of Tg.

For example, a configuration

would need 8.37×10^3 J for bond rotation, while a configuration

would need 62.8×10^3 J for rotation of the bond indicated.

(b) The presence of any cross-linking between polymer chains will stiffen the polymer, make bond rotation more difficult and, hence, raise Tg. Neutron irradiation can cause such cross-linking of amorphous polymers.

(c) The presence of plasticisers will usually lower Tg. Plasticisers are high molecular weight solvents for the polymer and tend to swell the polymer, holding the chain molecules further apart and giving a general increase in chain mobility. These plasticising liquids reduce the viscosity of the molten polymer and so increase the ease of fabrication and moulding. They themselves have Tg temperatures of 223–123 K and so can convert a rigid polymer to a rubbery material.

(d) Presence of co-polymer. The Tg temperature of a co-polymer is usually intermediate between those of the homopolymers, the actual temperature depending on the relative volume fractions of the homopolymers. The Tg temperature may thus be higher or lower than usual.

(e) Bulky side groups tend to stiffen the polymer chains and so raise Tg.

(f) An increase in symmetry will usually reduce Tg. Thus, PVC

has $Tg = 360$ K, while polyvinylidene chloride

has $Tg = 256$ K.

Apart from the drastic change in mechanical properties which occurs at Tg, there are also marked changes in properties such as volume, specific heat, coefficient of expansion and refractive index. It is obvious, therefore, that an amorphous polymer should not be operated near to its Tg temperature otherwise its behaviour will be unpredictable. Fortunately, there is a wide range of available materials, ranging from the vinyl polymers with Tg temperatures in the region of 470 K to the silicones with Tg temperatures as low as 150 K. Some typical Tg values are given in Table 5.1.

Table 5.1

Amorphous polymer	Repeat unit	Tg
Polyisoprene (rubber)		200 K
Polybutadiene (rubber)		183 K
PVC (rigid)		360 K
Polychloroprene (rubber)		223 K

These are all amorphous polymers and it is obvious that where Tg is below room temperature, the material will be a rubber. However, even natural rubber (polyisoprene) will become rigid if the temperature is lowered below 200 K.

2. Crystalline polymers (thermoplastics)

Crystalline polymers also have glass transition temperatures as seen in Table 5.2.

Table 5.2

Polymer	Repeat unit	T_g
Polyethylene	$\sim C - C \sim$ (with H, H above and H, H below)	153 K
Polypropylene	$\sim C - C \sim$ (with H, CH_3 above and H, H below)	263, 255 K
PTFE	$\sim C - C \sim$ (with F, F above and F, F below)	399 K

Fig. 5.2. Spherulites or crystalline regions in an amorphous background (courtesy Prof. Fraser P. Price, University of Massachusetts)

In many cases, the *Tg* temperature of a crystalline polymer is well below room temperature, as Table 5.2 shows, but yet the material is not rubbery. This is a basic difference between amorphous and crystalline polymers.

Whereas the properties of an amorphous polymer are a function of its *Tg* in relation to operating temperature, those of a crystalline polymer are mainly governed by the degree of crystallinity rather than by *Tg*.

No polymer is fully crystalline. The usual condition is for zones of crystallinity to exist, separated by amorphous or non-crystalline regions, as indicated in Fig. 5.2. The crystalline regions confer rigidity while the amorphous separating zones give toughness.

Crystallinity by definition is long-range ordering of atoms. This ordering can arise in polymers from three main sources. The simplest of these is the matching of parts of neighbouring chains, for example

A major part of the crystallinity is, however, the result of chains apparently folding back on themselves at intervals of about 100×10^{-10} m along their lengths. This produces matched, closely packed lamella. For example

A third way in which matching and long-range ordering may occur is by matching of the coils along a helical chain molecule, as shown by the example overleaf.

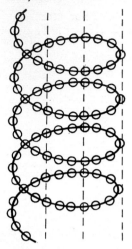

The crystalline zones are referred to as spherulites and since the presence of crystallinity increases mechanical strength and also softening temperature, it is desirable that these spherulites be as closely spaced as possible for maximum rigidity. This can be brought about by seeding the molten polymer with powdered solid polymer just before moulding and cooling. This gives a highly crystalline material which maintains its rigidity until close to its melting point.

Added crystallinity may also be induced by stretching the chains. This tends to untangle them, orientate them in the direction of stress and so produce more matching of chain segments. This drawing process is carried out at temperatures between Tg and the melting point and induces worthwhile extra strength. It is used extensively in the production of nylon and terylene filaments.

A plot of temperature against average molecular weight for crystalline polymers (Fig. 5.3) shows some differences from that for amorphous types.

In this case, a fairly definite melting temperature exists. It is also evident that crystallinity is not lost when the polymer is heated above Tg and, hence, crystalline polymers do not become rubbery above Tg.

The actual level of the Tg temperature in crystalline polymers is controlled by the same factors discussed for amorphous types, but since the overall properties are more a function of degree of crystallinity than of Tg, it is more important to understand those factors which influence crystallinity.

The ability of a polymer to crystallise is basically a function of the regularity of its molecular structure. Any highly random or irregular arrangement tends to produce amorphous types.

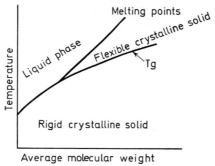

Fig. 5.3

Regularity of molecular structure is reduced by factors such as the following:

(*a*) Irregular substitution of foreign atoms into the polymer chains. Polyethylene is normally a highly crystalline rigid material, but if sulphur atoms are present during its manufacture, an amorphous, rubbery material may result.

(*b*) Co-polymerisation is very active in reducing crystallinity. This is to be expected since an ...AABABABBAB... type of chain molecule has much less chance of producing long-range ordering than a homopolymer chain. Co-polymerisation of crystalline polymers is often deliberately used to produce a more rubbery or flexible material, e.g. the random co-polymerisation of ethylene and propylene.

(*c*) Lack of stereoregularity reduces crystallinity. The bonds on carbon atoms are tetrahedrally spaced and so a chain molecule is not planar but more in the form of a zig-zag or helical arrangement. Because of this, different forms of the same molecule can exist, i.e. different isomers are possible, as follows

In this case all the X groups are regularly substituted. This is called the *isotactic* condition and is highly crystalline because matching across chains is highly probable.

(ii)

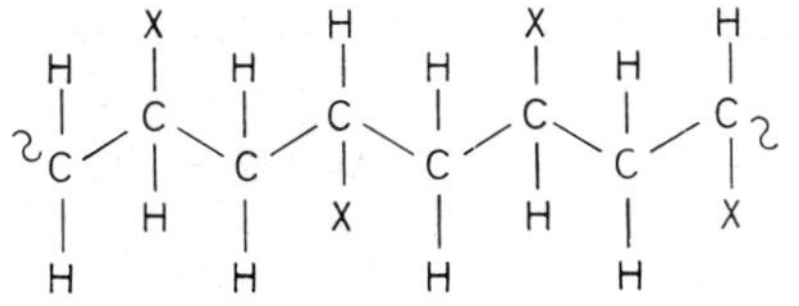

This form exhibits an alternation of the X groups but since the alternation is regular, there is still some ability to crystallise. This is the *syndiotactic* condition.

(iii)

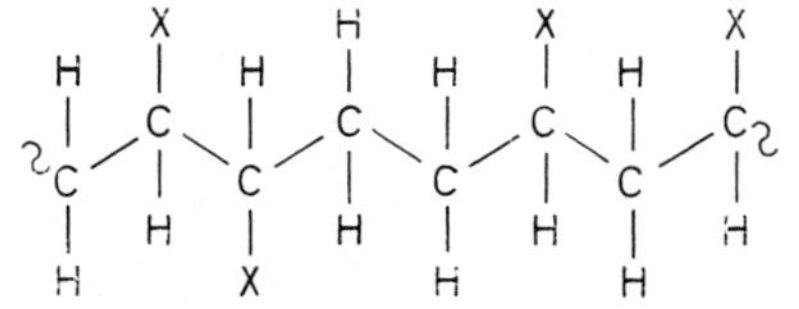

The substitution of X groups is completely random and the polymer will be amorphous. This is the *atactic* condition.

A degree of crystallinity over 80% is rare and most crystalline polymers exist in about the 50% crystalline 50% amorphous condition.

3. Cross-linked polymers (thermosetting)

In such materials, co-valent bonds occur between one chain and another instead of the usual van der Waals' bond. The degree of cross-linking will govern the final properties and, by controlling this factor, conditions varying from viscous liquid through to hard, rigid solid may be produced.

A typical lightly cross-linked material is engineering rubber. Natural rubber latex (polyisoprene) is a very elastic material.

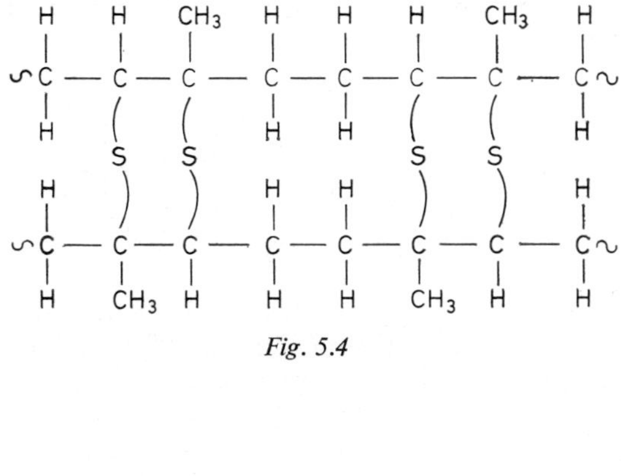

Fig. 5.4

It exhibits a large hysteresis in its response to stress and is only useful at normal temperatures. By vulcanising with sulphur, covalent cross-links are established giving more rigidity, more rapid response to stress variation and an extended operating temperature range. The cross-linking effect is illustrated in Fig. 5.4.

Cross-linking in rubbers may also occur as a result of oxidation, especially in the presence of strong sunlight, leading to hardening and eventual cracking, an effect often seen in old car tyres or windscreen sealing. As the degree of cross-linking increases, so will the hardness and rigidity. Ebonite, for instance, is simply rubber heavily cross-linked with sulphur. Heavy cross-linking is also found in the phenolic resins such as bakelite.

THE MECHANICAL BEHAVIOUR OF POLYMERS

Linear polymers consist of long-chain molecules with weak van der Waals' bonds between different chains. There is the possibility that these chains will slide past each other under stress so giving a viscous component to their elongation under stress. As well as this, there is also the chain stretching due to bond rotation and extension, this effect giving an elastic component. Polymers are not, therefore, perfectly elastic in their response to stress but are *viscoelastic* materials, i.e. they are partly elastic but they also have some of the characteristics of viscous liquids.

This sort of behaviour raises problems in mechanical testing and in design, since the mechanical properties and dimensions will be sensitive to the time over which the stress is applied and to the temperature of testing, because in a viscous or partly viscous material, the deformation is proportional to the viscosity of the material and the time over which the stress is applied, as well as being a function of the stress level. In an elastic material, of course, the deformation is a function only of the stress applied. Thus, in an elastic material,

$$\text{strain} \quad \varepsilon = \frac{\text{stress } \sigma}{\text{modulus } E}$$

and design parameters for a given temperature do not change with time, whereas in a viscous material,

$$\text{strain} \quad \varepsilon = \frac{\text{stress } \sigma}{\text{viscosity } \eta} \times \text{time } t$$

COLD DRAWING OF POLYMERS

If a polymer test piece is subject to a tensile test it may, at some stress, neck-down at some point on the gauge length. Continued straining then causes the necked region to spread along the gauge length until the whole sample has been reduced in diameter. During this drawing period, the load remains constant, indicating that the drawn region is carrying a higher stress than the undrawn parts. Such a drawing operation will therefore increase the strength and is used widely in the strengthening of polymer filaments. The elongations recorded during cold drawing may be as high as 500%, and so have little meaning for comparison purposes.

It is possible that the original necking-down of the sample generates sufficient heat to raise the temperature locally above Tg so allowing the molecular chains to slide past each other. The strengthening effect could then be equated with the increased ordering of the chains as they orientate in the direction of stress. Most polymers can be induced to draw if the right conditions of strain rate and temperature are chosen.

INFLUENCE OF TEMPERATURE ON STRENGTH

An amorphous polymer at temperatures above Tg will exist in a rubbery condition and will have a low modulus of elasticity. At temperatures well below Tg such a polymer will usually fail in a brittle manner and will have a much higher modulus. The effect of temperature variation on crystalline polymers depends on the degree of crystallinity. A moderately crystalline material (30–60% crystallinity) may be still quite tough at temperatures well below Tg, whereas a highly crystalline type (70–90%) could be brittle.

The variation of elastic modulus with temperature is characteristic of plastics and does raise problems in design. The change is typically from about 689·5 MN/m^2 at 373 K up to 6895 MN/m^2 at about 123 K. The modulus quoted for a room temperature test will obviously depend on the relative positions of Tg and room temperature.

The shape of a tensile stress–strain curve for a given polymer will also change drastically with temperature. Fig. 5.5 is the type of curve obtained at 293 K for Perspex and indicates a brittle material which fractures without measurable elongation.

Fig. 5.6 gives a curve for polypropylene at 293 K, indicating a tough material which necks-down and cold draws before fracture.

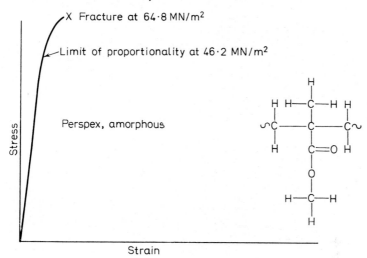

Fig. 5.5

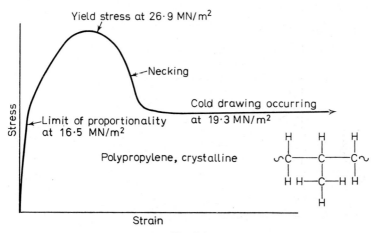

Fig. 5.6

Although these curves are for two different polymers, they could both well be for the same polymer at different test temperatures, i.e. brittle behaviour at low temperatures and tough behaviour at higher temperature. Engineering failure would be taken as the fracture point in the brittle condition or the yield point in the tough condition.

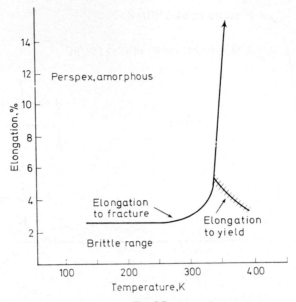

Fig. 5.7

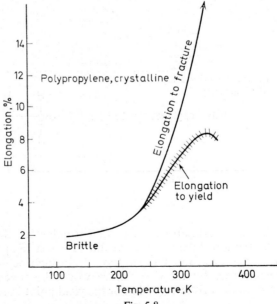

Fig. 5.8

The variation of tensile elongation with temperature is indicated in Figs 5.7 and 5.8. In the brittle condition, the elongation needed to cause fracture is less than the elongation to cause yielding. At some temperature, this position is reversed and the material becomes tough. The cross-over point on the curves is roughly equatable to Tg in both cases.

The decrease in elongation to yield as temperature increases is typical of amorphous polymers. Crystalline polymers, as Fig. 5.8 indicates, behave in the opposite way.

INFLUENCE OF STRAIN RATE ON POLYMER STRENGTH

Movement of and within polymer chains is subject to hysteresis, i.e. the deformation lags behind the application of stress. The actual deformation D_T at any time T is related to the theoretical deformation D_A via a relationship of the type

$$D_T = D_A(1 - e^{-T/\tau_m})$$

where τ_m is the orientation time of the molecular chains. If T is much smaller than τ_m, then in the time available very little deformation will occur and the material will behave as a brittle solid. Application of stress over longer periods of time would cause the polymer to behave as either a tough or a rubbery material.

In general, increasing the strain rate will increase the yield stress until it becomes higher than the fracture stress and brittle failure will occur. Shock loading of this nature is, of course, induced by the presence of stress concentrations or notches. Table 5.3 gives some comparisons between the properties of polymers and other materials.

Table 5.3

Material	Density kg/m³	Thermal conductivity W/m deg C	Coefficient of expansion deg C⁻¹	Resistivity Ωm	Modulus MN/m²
Aluminium	$2 \cdot 7 \times 10^3$	167·5	$21 \cdot 6 \times 10^{-6}$	$3 \cdot 5 \times 10^{-8}$	6 895
Steel	$7 \cdot 85 \times 10^3$	46·1	$11 \cdot 3 \times 10^{-6}$	17×10^{-8}	20 685
Phenol form-aldehyde	$1 \cdot 3 \times 10^3$	0·17	72×10^{-6}	1×10^{10}	3 447
Polyethylene	$0 \cdot 9 \times 10^3$	0·33	180×10^{-6}	1×10^{11}	
PTFE	$2 \cdot 2 \times 10^3$	0·21	99×10^{-6}	1×10^{14}	
Nylon	$1 \cdot 15 \times 10^3$	0·25	99×10^{-6}	1×10^{12}	2 758

The table indicates polymers as being much lighter than metals. The bonding mode between polymer chains is reflected in their low elastic moduli as compared with metals. It is also indicated that the co-valent bonding within the molecules of a polymer produces a material which is basically a good electrical insulator.

VISCOELASTICITY IN POLYMERS

A viscoelastic material is one which will change its dimensions slowly under stress over long periods of time, even though these stresses may well be below those which would normally cause permanent deformation. This *creep* is more pronounced at elevated temperatures and must be allowed for in design.

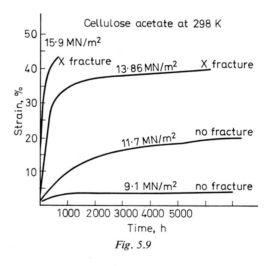

Fig. 5.9

Creep behaviour is usually investigated by measuring the variation of strain with time at constant temperature and load. Plots of strain–time are of characteristic shape, as shown in Fig. 5.9. The curves indicate:

(*a*) an instantaneous elongation immediately on application of load—this deformation is inversely proportional to the elastic modulus ($\varepsilon = 1/E$) and so is purely elastic;
(*b*) this is followed by a rapid rate of creep which leads into
(*c*) a constant rate of creep which may approach zero.

It is also possible to plot creep compliance–time, i.e.

$$\text{compliance} = \frac{\text{elongation at time } t}{\text{stress at time } t} = \frac{\text{strain}}{\text{stress}}$$

Since modulus is σ/ε, the compliance may be regarded as a reciprocal modulus which changes with time. For materials in which this compliance does not change as load changes, the creep curves become a single curve for all stresses, i.e. all the curves above would plot on to a single curve.

A mechanical model, consisting of elastic springs and dashpots containing a Newtonian fluid, illustrates some of the points regarding creep in viscoelastic material. Fig. 5.10 illustrates this model.

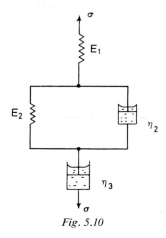

Fig. 5.10

The strain in elastic springs E_1, E_2 is a function only of stress. The strain in dashpots η_2, η_3 which contain a perfect viscous liquid is proportional to time and stress and also to $1/\eta$ of the liquid, where η is the viscosity.

Consider a stress σ applied to the model. The total stress is carried by both E_1 and η_3. The total stress is shared between E_2 and η_2, equally at first but, as the dashpot η_2 extends, more and more of the stress will be carried by E_2. At some point, all the stress will be carried by E_2 and none by η_2.

The elongation of this parallel part of the model is given by

$$\varepsilon_2 = \frac{\sigma}{E_2}(1 - e^{-t/\tau})$$

where t is the time and τ is the retardation time $= \eta_2/E_2$. This retardation time is used as a matter of convenience and is that time at which the parallel combination would deform by $(1-1/e) = 63 \cdot 21\%$ of its total deformation. The time for total deformation would be impossibly long and so τ is used as a practicable period.

The total deformation of the whole model is then deformation of E_1+ that of η_3+ that of parallel combination, i.e.

$$\varepsilon = \frac{\sigma}{E_1} + \frac{\sigma}{\eta_3} \cdot t + \frac{\sigma}{E_2} (1 - e^{-t/\tau})$$

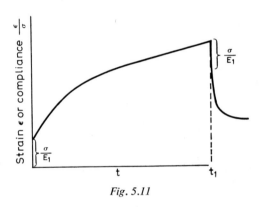

Fig. 5.11

The model produces a creep curve of the type illustrated in Fig. 5.11. On applying the stress to the model:

(a) E_1 at once stretches by σ/E_1, the initial elastic extension.
(b) Creep begins at a high rate as the dashpots extend, but as this extension lowers the stress on the dashpot the creep rate becomes less and reaches a constant value.
(c) After time t_1, load is removed. E_1 contracts elastically by σ/E_1.
(d) Recovery then begins as E_2 gradually forces dashpot η_2 towards its original condition. However, η_3 does not fully retract since there is now no force on it, hence the recovery curve flattens out.

The creep curve so obtained is often re-plotted as log ε/log t, as shown in Fig. 5.12.

On such plots, the steepest part of the creep curve occurs at a time equal to τ. Most of the creep, in a model such as this, occurs in a period of time about one decade around the retardation time. The recovery curve is almost a mirror image of the creep curve and,

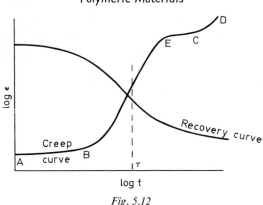

Fig. 5.12

again, most recovery occurs about one decade around the retardation time.

This sort of model illustrates many of the features of creep curves but it can only give a qualitative picture. The strain ε_2 represents uncoiling and untangling of the chain molecules in a real polymer. Obviously, because of the random way in which the molecules have grown, different degrees of entanglement will exist and thus different parts of any given molecule will have different retardation times.

Real materials, therefore, differ from the model in two major ways: (*a*) creep takes place over many decades of time and not during just one decade; (*b*) many separate retardation times exist distributed over many decades of time.

INFLUENCE OF Tg AND TEMPERATURE ON CREEP (AMORPHOUS POLYMERS)

Creep is a temperature-dependent phenomenon and we would expect creep behaviour, particularly of amorphous polymers, to undergo a considerable change in the region of *Tg*.

At temperatures well below *Tg* not much creep will occur, since segmental movement of the polymer chains is restricted, i.e. the viscous component of the deformation is due to the movement of parts of chains relative to each other.

Raising the temperature causes an increase in both creep rate and elongation, since segmental motion becomes possible.

In the region of *Tg*, creep becomes very sensitive to temperature and there is a drastic change in modulus *E* (modulus decreases).

For example,

tensile compliance well below $Tg = \varepsilon/\sigma \simeq 3 \times 10^{-12}$ N/m²

tensile compliance around $Tg = \varepsilon/\sigma \simeq 1 \times 10^{-10}$ N/m²

Thus, since $\varepsilon = \sigma \times$ compliance, any given load applied at Tg will give much greater elongations than the same load well below Tg.

It is also observed that the change in ε with time undergoes a marked change around Tg, i.e. the creep rate goes through a maximum near Tg. At temperatures well above Tg, the creep *rate* usually gets less even though total elongations may be greater. These points are illustrated in Fig. 5.13.

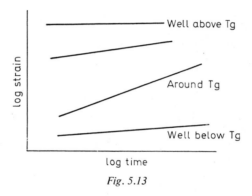

Fig. 5.13

Materials which give straight-line relationship between log compliance and log time obey an equation of type $\varepsilon = K\sigma t^n$, i.e.

$$\log \varepsilon = \log K + \log \sigma + n \log t$$

where K, n are constants.

With $n = 0$, the material would be perfectly elastic since then ε is proportional to σ. With $n = 1$, the material is a perfect viscous fluid. Thus n measures the relative contributions of elastic and viscous contributions to creep deformation.

INFLUENCE OF MOLECULAR WEIGHT ON CREEP

With amorphous polymers, the molecular weight has very little influence on creep at temperatures well below Tg. Creep involves mainly the motion of chain segments relative to each other and the total length of the chain (which is the molecular weight) is not involved.

Above Tg, amorphous polymers behave as rubbers and whole chains may move relative to each other. Creep thus becomes dependent on molecular weight. The creep rate is governed mainly by the viscosity of the polymer in this condition (the higher the viscosity the lower the creep rate) and viscosity increases as molecular weight increases. The presence of plasticisers will also reduce viscosity and so such polymers may exhibit high creep rates.

As a general summary, therefore, of creep in amorphous polymers:

(a) Such materials have creep curves similar to the $\log \varepsilon/\log t$ curve plotted for the model but the real creep curve is spread over a much wider time scale,

(b) At temperatures well below Tg only the first part $A–B$ of the $\log \varepsilon/\log t$ plot would be observed (i.e. before the curve climbs),

(c) At temperatures well above Tg, only the part $C–D$ would be recorded,

(d) Around Tg, the whole curve would be recorded in a short time,

(e) As molecular weight increases, the plateau $E–C$ becomes more pronounced and may extend over very long periods of time. This plateau is the result of chain entanglements and these give a certain amount of elastic behaviour to the polymer and so delay the onset of the stage $C–D$ which occurs when viscous deformation begins to dominate the creep process.

CREEP IN CROSS-LINKED POLYMERS

The degree of cross-linking has little influence on creep well below Tg since the polymer is so rigid in any case.

In lightly cross-linked polymers, such as vulcanised rubbers, the presence of cross-links will radically influence creep above Tg. Cross-links are chain entanglements and these give elastic response to stresses. Ideally, therefore, the presence of cross-linking should produce a perfectly elastic material above Tg instead of one which creeps (viscous component). In practice, this aim is never fully realised and all rubbers exhibit some degree of continuing extension under maintained loading.

Obviously, if the polymer is very heavily cross-linked, the materials become rigid with Tg above decomposition temperature. Such materials are not subject to creep. The effect of cross-linking is indicated in Fig. 5.14, which refers to rubber.

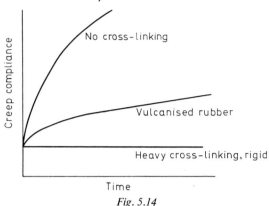

Fig. 5.14

CREEP IN CRYSTALLINE POLYMERS

With low degrees of crystallinity, the crystalline regions act as cross-links. The creep behaviour is thus as before, i.e. little creep below Tg, some creep above Tg, depending on degree of cross-linking. Plasticised PVC is typical in this respect. It is rubbery but has good dimensional stability. This behaviour goes on up to about 20% crystallinity. Above about 40% crystallinity the crystalline regions begin to form a continuous network throughout the polymer and so they (rather than the amorphous regions) begin to carry the majority of any applied stress. This results in a drastic reduction of creep compliance (increased modulus).

As we have seen, the creep properties of amorphous or lightly cross-linked polymers are not very sensitive to temperature changes provided the temperature is well below Tg. In contrast to this, the creep compliance of crystalline polymers is rather sensitive to temperature, mainly because the degree of crystallinity changes as temperature changes. However, a crystalline polymer creeps very little at any temperature when compared with the same material in the amorphous state.

STRESS RELAXATION IN POLYMERS

Stress relaxation is another aspect of viscoelasticity. In this sort of test, a stress is applied instantaneously to produce a definite strain. The stress needed to maintain this strain decreases with time and this relaxing value of stress is measured.

The measurements yield similar information to that given by

creep tests but results are expressed in different ways. For small extensions and slow relaxation, creep and stress relaxation results are comparable and the stress relaxation curve is a mirror image of the creep curve. The relationship is less simple at high creep strains and high rates of stress relaxation.

Stress-relaxation phenomena are observed in any viscoelastic material, whether plastics, ceramics or metals. Thus it may be necessary to determine the stress needed to hold a metal insert in a plastic component over a long period of time, to determine the stress needed to maintain tightness in a high temperature bolted assembly or to determine the time and temperature needed to relieve residual stresses in a metal.

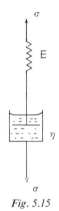

Fig. 5.15

The simplest model used to illustrate stress relaxation behaviour is the Maxwell model, consisting of a spring and a dashpot in series, as shown in Fig. 5.15. If a stress is applied to a Maxwell model, the equation of motion is

$$\frac{d\varepsilon}{dt} = \frac{\sigma}{\eta} + \frac{1}{E} \cdot \frac{d\sigma}{dt}$$

Since the elongation ε is not allowed to change with time, $d\varepsilon/dt = 0$ and so the equation can be integrated to give

$$\sigma = \sigma_0 e^{-(Et/\eta)} = \sigma_0 e^{-t/\tau}$$

where σ_0 is the initial applied stress; σ is the relaxed value of stress at any time t; $\tau = 1/E$, which is the relaxation time. This is similar to the retardation time in creep and is chosen as the time required for the stress to decay to $1/e$ or $36\cdot79\%$ of its original value. It is chosen to be this value partly because the plot of σ/t has maximum slope at this time level.

Since
$$\sigma = \sigma_0 e^{-t/\tau}$$
$$\log \sigma = \log \sigma_0 - \log_{10} e \times \frac{t}{\tau}$$

so
$$\log \sigma = \log \sigma_0 - \frac{t}{2 \cdot 3\tau}$$

In the model, all the initial deformation goes into stretching the spring (the elastic component) but this at once exerts a force on the dashpot (viscous component) which stretches and so relaxes the stress. Over very long periods, the stress would eventually decay to zero but the times involved are so long that the concept of τ is used.

The results of stress relaxation tests on such a model give characteristic curves, as illustrated in Fig. 5.16.

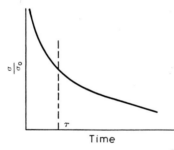

Fig. 5.16

On the model, most of the relaxation takes place within one decade of time around the relaxation time. The curves can also be used to give an idea of the variation of the modulus E of a viscoelastic material with time, i.e.

$$\frac{\sigma}{\varepsilon} = E = \frac{\sigma_0}{\varepsilon} \cdot e^{-t/\tau}$$

Thus, as stress relaxation occurs, ε remains constant but σ_0 decreases and, hence, the modulus decreases with time.

The model is useful in analysing stress relaxation behaviour but it can give only a qualitative picture since real polymers have more than one relaxation time. This is to be expected since polymer chains have varying degrees of entanglement with each other and so each chain will have variable elastic and viscous components along its length. There is therefore an infinite number of relaxation times in real polymers and these cover many decades of time. Hence the $(\sigma/\sigma_0)/\log t$ plot for a real polymer would be much flatter than the model curve and would cover a much greater time scale. In such cases it is necessary to evaluate the distribution of relaxation times.

More homogeneous materials such as metals tend to have finite relaxation times and so the relaxation equations can be used direct.

As regards the stress relaxation behaviour of real polymers, the following points are of interest.

Influence of temperature and transitions (amorphous polymers)

As would be expected, stress relaxation behaviour is very sensitive to the temperature of stressing since higher temperatures tend to accentuate the viscous component. This sensitivity to temperature becomes very marked in the case of amorphous polymers in the Tg region, i.e. the polymer is passing into a rubbery state. The greatest sensitivity occurs while the transition is actually occurring —once the change is completed and the material is a rubber, the

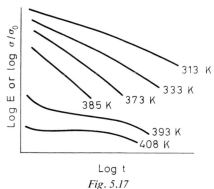

Fig. 5.17

sensitivity becomes less pronounced. This is illustrated in Fig. 5.17, which is a plot of $\log E/\log t$ for Perspex. $Tg = 385$ K (since ε is constant, $\log E$ is really $\log \sigma_0$).

Influence of molecular weight and cross-linking (amorphous polymers)

Increasing either molecular weight or degree of cross-linking gives added stiffness to an amorphous polymer and the polymer will hold stress for much longer periods of time. Increasing the molecular weight is particularly beneficial at temperatures above Tg.

Cross-linking is mainly present in rubbers—these are polymers whose Tg temperatures are below room temperature. The cross-links establish co-valent bonds between chains and these (*a*) will prevent the stress from relaxing to zero, (*b*) will prevent the stretched-out molecular segments from coiling back again to their original condition as the stress relaxes.

Thus, there is a tendency with rubbers to reach an equilibrium value of modulus E during a stress relaxation test and this value of E can be estimated as

$$E = \frac{3RTd}{M_c}$$

where R is the gas constant; T is absolute temperature; d is density; M_c is number average molecular weight.

Such a relationship indicates an *increase* of modulus as temperature increases and, hence, cross-linked polymers behave differently from normal amorphous types. This effect is a result of the increased thermal vibration causing the molecular segments to pull more strongly at their anchor points and try to coil up more closely. (One could perhaps visualise the modulus of a vibrating spring increasing with the frequency of vibration.)

The presence of an equilibrium value of modulus E will tend to produce a plateau on the modulus–time curve. For example, Fig. 5.17 shows some indication of this, and this effect would be even more pronounced in a truly cross-linked polymer. The plot for Perspex in Fig. 5.17 shows a plateau above Tg because the normal chain entanglements are acting as temporary cross-links.

Influence of crystallinity (crystalline polymers)

Spherulites or crystal regions in a polymer act in the same way as chain entanglements and so each part of a chain molecule will have its own relaxation time. Thus the main influence of crystallinity is to widen the distribution of relaxation times and extend the relaxation of stress to much longer times. Log E/log t curves for crystalline polymers tend to be much flatter than those previously given for amorphous types. This effect is true at all temperatures and, as usual, stress relaxation in such polymers is not a function of Tg in relation to operating temperature but a function of degree of crystallinity. The modulus, as usual, decreases as temperature increases.

The phenomenon of drawing has been mentioned, i.e. the material elongates at constant load thus indicating that stress-induced ordering or crystallisation is producing a strengthening effect. In a stress relaxation test, if the applied stress is near to the drawing stress and drawing sets in, then obviously the stress needed to maintain a constant strain will rapidly decrease. This stress-induced crystallisation in polymers will cause rapid stress relaxation.

Stress relaxation tests are very useful in studying the degradation of polymers such as rubbers since the relaxation behaviour is very sensitive to the degree of cross-linking. Hence the influence of chemical reactions in rubbers can be studied by stress relaxation techniques.

STABILITY OF POLYMERS

Polymers are usually considered to be rather inert chemically but they are prone to chemical attack in some environments. The resistance to chemical attack is only as good as the weakest bond in the molecular structure and so some polymers will be more resistant than others.

Double bonds between carbon atoms such as occur in natural and synthetic rubbers are inherently unsatisfied and are capable of being broken by active elements such as oxygen and chlorine.

The simple olefine polymers such as polyethylene and polypropylene contain only C—C and C—H bonds and are really paraffins of very high molecular weight. As such they tend to be inert but again they may react with active elements such as the halogens.

Polymers which inherently contain C-halogen bonds are usually very stable and chemically inert. This is typical of polytetrafluorethylene

$$
\begin{array}{ccc}
\text{F} & \text{F} & \text{F} \\
| & | & | \\
\sim\!\text{C} - \text{C} - \text{C}\!\sim \\
| & | & | \\
\text{F} & \text{F} & \text{F}
\end{array}
$$

Although polyvinyl chloride

$$
\begin{array}{cc}
\text{H} & \text{H} \\
| & | \\
\sim\!\text{C} - \text{C}\!\sim \\
| & | \\
\text{H} & \text{Cl}
\end{array}
$$

also contains C-halogen bonds, its chemical reactivity is much higher than that of PTFE. This is mainly the result of weak spots in the polymer chains which act as preferred sites for initiation of chemical reactions.

Most polymers contain such active sites and these are often at the chain ends. If a chain end is attacked in this way, the end group is removed so exposing a mer, now with an unsatisfied bond. This likewise is removed and eventually the whole chain may be unbuilt, changing the material from a rigid high molecular weight condition into a rubbery or viscous liquid condition. Degradation of this type may be prevented by removing reaction sites. Such sites within the chains may be prevented by introduction of co-polymers which act as blocking points along the chain. Chain end stability is usually achieved by capping the ends with a very stable terminating agent.

Chemical attack could also induce cross-linking by opening out any double bonds. In either case, a marked change in mechanical properties would occur.

Dissociation of bonds in polymer molecules can also occur as a result of absorption of radiant energy, particularly in the presence of oxygen or halogen elements. Again, the influence of radiant energy on a polymer is a function of the bond strength.

Table 5.4 indicates the result of subjecting polymers to thermal energy *in vacuo*. The rate of degradation of the polymer at 623 K has been measured. The high stability of the C–F bond in PTFE is again evident.

Table 5.4

	Polymer	*Rate of degradation* %
PTFE	$\begin{array}{ccc} F & F & F \\ | & | & | \\ \sim C & - C & - C \sim \\ | & | & | \\ F & F & F \end{array}$	0·000005
Perspex	$\begin{array}{ccc} H & CH_3 & H \\ | & | & | \\ \sim C & - C & - C \sim \\ | & | & | \\ H & C{=}O & H \\ & | & \\ & O & \\ & | & \\ & CH_3 & \end{array}$	5·2
PVC	$\begin{array}{cc} H & H \\ | & | \\ \sim C & - C \sim \\ | & | \\ H & Cl \end{array}$	170

Polymers may also be affected by light, particularly by ultra-violet radiations and strong sunlight, which may either cause cross-linking or degradation. Short wavelength radiation carries enough energy to rupture some of the bonds in polymers but damage is not always caused, since it is found that any given bond will only absorb energy at definite frequencies. Resistance to damage by short wavelength radiation may be improved by adding substances which will preferentially absorb damaging radiations. For example, the addition of carbon black to rubbers reduces their hardening by ageing.

High energy radiations such as γ- or X-radiation may also influence the properties of polymers and again may either cause degradation or cross-linking depending on the conditions. Polyethylene, for example, can be improved by deliberate irradiation. A C–H bond is broken leaving a free C–bond which then co-valently cross-links into another chain. For example,

The net result is a more rigid material with a higher than usual softening temperature. Similar stiffening effects are found in rubbers and polypropylene.

Generally, however, high energy radiation will cause degradation and softening of polymers and those which initially cross-link will degrade and soften if given an overdose of radiation.

PROPERTIES AND APPLICATIONS OF TYPICAL ENGINEERING POLYMERS

It is not feasible, in a book of this nature, to cover in detail the properties of all available polymers. It is therefore proposed to deal with a few types only in order to illustrate the general principles of manufacture and to indicate the properties available.

Polyethylene (crystalline thermoplastic)

The mer is the gas ethylene

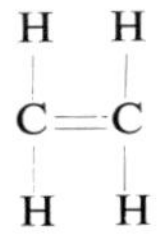

$$\begin{matrix} H & & H \\ | & & | \\ C & = & C \\ | & & | \\ H & & H \end{matrix}$$

available as a by-product from petroleum refining. The polymerisation is an addition reaction, the double bond being induced to open out and add on to similarly treated mers.

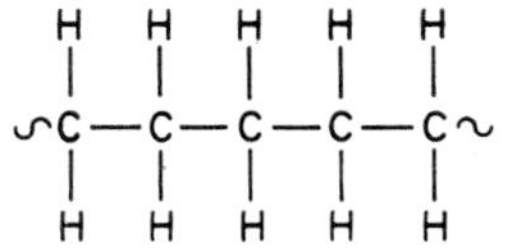

$$\begin{matrix} H & H & H & H & H \\ | & | & | & | & | \\ \curlyvee C - C - C - C - C \curlyvee \\ | & | & | & | & | \\ H & H & H & H & H \end{matrix}$$

The original polymerisation process involved the use of high pressures and temperatures in the presence of an initiator such as benzoyl peroxide. The benzoyl peroxide, like all initiators, provides radicals with free or available bonds for starting off the addition polymerisation process, and also catalyses the breakdown of the double bonds in the mers.

Thus benzoyl peroxide decomposes into free radicals:

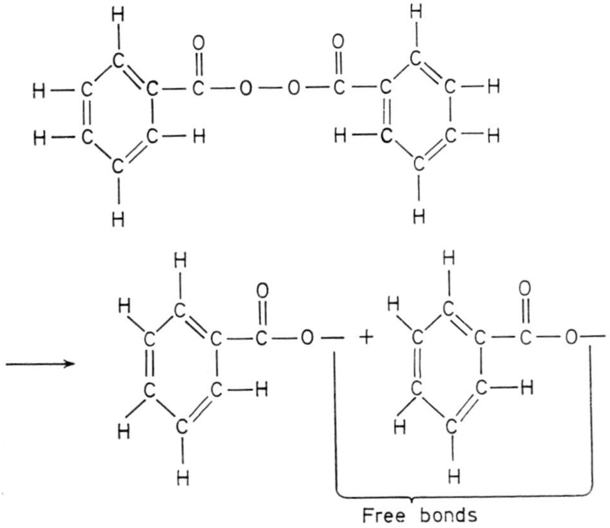

Free bonds

or more simply   R—R = R—+R—

The free radical R— attaches itself to one end of a mer leaving an unsatisfied bond at the other end, i.e.

$$R-\overset{\displaystyle H}{\underset{\displaystyle H}{C}}-\overset{\displaystyle H}{\underset{\displaystyle H}{C}}-$$

Further mers, whose double bonds have been opened, then add on to the chain to build the polymer molecule.

It is quite likely that the initiator will also stop the growth of a polymer chain by capping the other end, i.e.

$$R-\overset{H}{\underset{H}{C}}-\overset{H}{\underset{H}{C}}-\overset{H}{\underset{H}{C}}-----\overset{H}{\underset{H}{C}}-R$$

and it is the resultant random length of the individual chains which makes the use of an average molecular weight necessary.

This type of 'high pressure' polyethylene tends to consist of branched chains and the entanglement between molecules which results from branching reduces the degree of crystallinity to about 50%. As a result, this material is of comparatively low density (0.92×10^3 kg/m³) and of low melting point (388 K).

A higher density form (0.96×10^3 kg/m³) with a melting point of 408 K can be produced by treating a solution of ethylene in paraffin at 403–433 K and 1.38–3.5 MN/m² pressure in the presence of a SiO_2–Cr_2O_3 catalyst. This material has substantially unbranched molecules and may reach over 90% crystallinity.

The general properties of the two types of polyethylene are indicated in Table 5.5. The low density material quoted in the table will have about 20 CH_3 groups in a chain 1000 carbon atoms

Table 5.5

Property	Low density material	High density material
Max stress, MN/m²	15·2	27·6
% Elongation to fracture	620	500
Softening temperature, K	371	
Melting point, K	381	408

long. High density material has about 1·5 CH_3 groups per 1000 carbon atoms. Since a CH_3 group may represent a chain ending, i.e.

$$\sim C-C-C-H$$

it is obvious that the high density material will have the highest average molecular weight.

In the production of this, as with other polymers, it is usual to blend into it additives to confer desired properties. These additives may include: (*a*) pigments for colouring; (*b*) rubbers to give extra flexibility; (*c*) lubricating agents to aid processing; (*d*) cross-linking agents to give extra rigidity for elevated temperature operation; (*e*) antistatic agents to reduce dust attraction; (*f*) antioxidants to retard hardening during ageing.

Fabrication of the polymer, as with most thermoplastics, involves heating near to or just above the melting point and then shaping by extrusion or by pressure moulding.

The mechanical properties of polyethylene, as one would expect, depend on such factors as temperature, strain rate, density, degree of crystallinity. By increasing density and strain rate, or by decreasing the temperature, the modulus of elasticity is increased. For example, a low density material would give a modulus of $138\,MN/m^2$, whereas high density material might have a modulus as high as $689\,MN/m^2$.

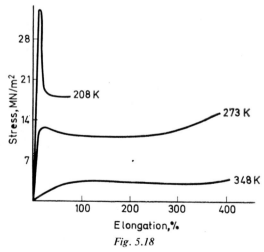

Fig. 5.18

Table 5.6

Strain rate m/s	Max. stress MN/m²	Elongation to fracture %
0·0025	18·5	380
0·0075	20·0	200
0·0125	22·1	180

Fig. 5.18 indicates the influence of temperature on tensile properties, while Table 5.6 illustrates the influence of strain rate.

Polyethylene is fairly inert chemically but it does age on exposure to air and sunlight and this ageing results in a loss of toughness and electrical insulating properties. The material is also prone to environmental stress cracking, i.e. it will fail in a brittle manner if it is stressed in the presence of substances such as soaps, detergents and alcohols. Resistance to this type of embrittlement is improved by using high density material or by co-polymerising with about 5% of butylene.

The vast majority of the polyethylene produced is taken for packaging but about 10% of the output is used for electrical insulation. It is also being used in the form of tubing for water supply and as a surface coating on metallic materials.

Polypropylene (crystalline thermoplastic)

The monomer propylene

$$\begin{array}{cc} H & H \\ | & | \\ C = C \\ | & | \\ H & H-C-H \\ & | \\ & H \end{array}$$

obtained from petroleum distillation is heated under pressure in the presence of a chloride catalyst. The polymerisation temperature is below the melting point of the polymer and so polypropylene precipitates as a slurry, i.e.

$$\begin{array}{ccccc} H & H & H & H & H \\ | & | & | & | & | \\ \sim C - C - C - C - C \sim \\ | & | & | & | & | \\ H & CH_3 & H & CH_3 & H \end{array}$$

Isomerism occurs in this structure and can lead to:

(*a*) the isotactic form, as indicated above, with all the CH_3 groups regularly disposed on the same side of the molecule;

(*b*) the atactic form

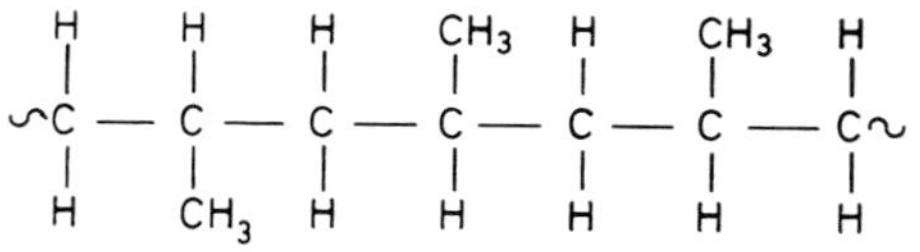

and

(*c*) the syndiotactic form

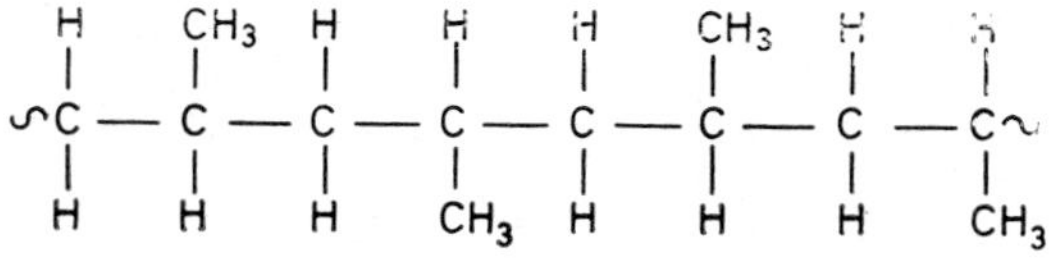

The atactic and syndiotactic forms have the CH_3 groups on either side of the chains and matching between different molecules is not possible. These are, therefore, amorphous. The isotactic condition is desirable and the polymer slurry is usually treated with selective solvents to remove the amorphous forms. Commercial polymers will, however, contain some of the amorphous forms and their presence reduces the degree of crystallinity and hence influences the mechanical properties, as shown in Fig. 5.19.

Polypropylene is one of the lightest plastics (sp.gr. $= 0.905 \times 10^3$ kg/m^3). It is usually highly crystalline being about 90–95% isotactic and this, together with the relative bulkiness of the CH_3 side-groups, gives it good stiffness and a softening temperature above 373 K.

The homopolymer has the disadvantage of becoming brittle at around 273 K but this fault is usually suppressed by co-polymerisation with a few per cent of ethylene.

Polypropylene tends to be chemically more active than polyethylene since the CH_3 side-groups are more open to attack. Whereas polyethylene cross-links on oxidation and becomes brittle, polypropylene will degrade into smaller molecules.

The polymer is widely used in place of polyethylene because of its better resistance to softening at temperatures in the region of 373 K, e.g. as mouldings for auto components and washing machine parts. The polymer is also used as cold-drawn filament in the production of ropes, nets and woven fabrics.

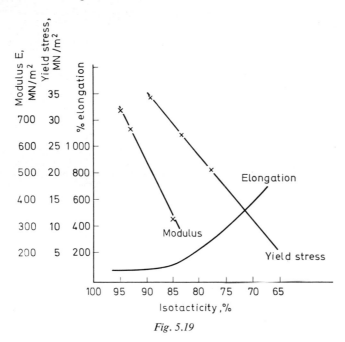

Fig. 5.19

Rubbers (amorphous or partially cross-linked polymers)

Most rubbers are derived from a group of organic molecules called *dienes*. They are polymers but, because of their rubber-like nature, they are often referred to as elastomers. The vast majority of rubber used in engineering is synthetic rubber obtained by polymerising petroleum by-products.

The molecules used for rubber formation are based on the diene arrangement, i.e.

where the side-group R can be H giving butadiene, Cl giving chloroprene, CH_3 giving isoprene.

Natural rubber latex is a polymer of isoprene. These structures exhibit isomerism since the unsaturated bonds can occur in different parts of the molecule. For example, natural rubber latex has a structure

and is referred to as *cis*-1, 4 polyisoprene, the word '*cis*' meaning same side, i.e. the unsaturated carbon bonds lie on the same side of the molecule at the 1,4 positions. Another possible arrangement is

This is the '*trans*' isomer since it has unsatisfied bonds on opposite sides of the molecule. This is actually the molecule of guttapercha which is anything but rubber-like in properties.

Natural rubber

This is a high molecular weight polymer of isoprene in which the *cis*-1,4 arrangement predominates. Like all rubbers based on dienes, the polymer chain will contain one residual double bond per repeat unit and this can be opened out to produce cross-linking. The material is amorphous in the unstrained condition and obviously operates above its glass transition temperature.

Controlled cross-linking or vulcanisation of natural rubber has been outlined on p. 182. The establishment of co-valent bonds between the chains gives greater rigidity and better strength and elasticity. Some 3–5% of sulphur is used.

Natural rubber, when strained, undergoes re-orientation of its molecules and the matching produces some crystallinity. The material thus becomes noticeably harder and stronger under stress,

and at high strain rates it may even appear to be brittle. The increased hardness under load is useful in providing wear resistance and
in giving rapid response to load changes.

Butadiene rubbers

These are synthetic rubbers using butadiene obtained from petroleum distillation. The monomer is polymerised using metallic
sodium Na as an initiator, hence the name BUNA rubbers.
The polymer arrangement is

$$\sim C - C = C - C - C - C = C - C \sim$$

The raw polymer is improved considerably by controlled cross-
linking using sulphur or oxygen via the residual double bond
positions. Further improvement of properties can be given by co-
polymerising with styrene and this produces SBR, the major general-
purpose synthetic rubber.
Styrene is a stiff rigid polymer of structure

and this, when distributed randomly into the polybutadiene chains,
gives added stiffness to the rubber. The random distribution,
however, means that the material will not crystallise under stress
as will natural rubber, and so for tyre manufacture a filler is needed
to give abrasion resistance.

Nitrile rubbers

These are produced by co-polymerising butadiene

$$\begin{array}{cccc} H & H & H & H \\ | & | & | & | \\ C = C & - & C = C \\ | & & & | \\ H & & & H \end{array}$$

and acrylonitrile

$$\begin{array}{cc} H & H \\ | & | \\ C = C \\ | & | \\ H & C \equiv N \end{array}$$

The material is very resistant to petroleum products and widely used for petrol hoses and collapsible fuel tanks. It may also be used as a filler in thermosetting polymers to give better toughness.

Butyl rubbers

These are co-polymers of isobutylene

$$\begin{array}{cc} & H \\ & | \\ H & H-C-H \\ | & | \\ C === C \\ | & | \\ H & H-C-H \\ & | \\ & H \end{array}$$

and isoprene

$$\begin{array}{cccc} & H \\ & | \\ H & H-C-H & H & H \\ | & | & | & | \\ C === C & - & C = C \\ | & & | \\ H & & H \end{array}$$

produced by mixing liquid isobutylene with about 5% of liquid isoprene and polymerising at about 180 K in the presence of a chloride catalyst. The solid polymer precipitates out as a slurry.

Like natural rubber, butyl rubbers are normally amorphous but become crystalline to some extent when strained. Unlike natural rubber, however, butyl rubbers do not crystallise on cooling and so will remain elastic down to temperatures as low as 220 K. As with most rubbers, cross-linking is necessary to give controlled properties. Butyl rubber exhibits pronounced hysteresis in its elastic response to stress, and deformation lags well behind stress oscillation indicating a high internal friction and damping capacity. This sluggishness in elastic response is useful in improving road-holding in tyres but it also means that the tyres overheat more rapidly than usual. Butyl rubbers have very low permeability for gases and so are widely used for tyre inner tubes.

Polymethylmethacrylate—Perspex (linear amorphous thermoplastic)

This is one of the family of acrylic plastics based on acrylic acid

$$
\begin{array}{ccc}
\text{H} & & \text{H} \\
| & & | \\
\text{C} & = & \text{C} \\
| & & | \\
\text{H} & & \text{C}-\text{O}-\text{H} \\
& & \| \\
& & \text{O}
\end{array}
$$

The mer of Perspex is methylmethacrylate

$$
\begin{array}{ccccc}
\text{H} & \text{CH}_3 & \text{O} & & \text{H} \\
| & | & \| & & | \\
\text{C} & = \text{C} & - \;\; \text{C}-\text{O}- & \text{C} & -\text{H} \\
| & & & | \\
\text{H} & & & \text{H}
\end{array}
$$

Polymerisation occurs easily at about 373 K with a free radical initiator. The polymer precipitates out as an insoluble solid and is produced by opening out of the double bond between the carbon atoms to give

$$
\begin{array}{c}
\text{H} \\
| \\
\text{H}\;\;\text{H}-\text{C}-\text{H}\;\;\text{H} \\
|\qquad|\qquad| \\
\sim\text{C}-\!-\!-\text{C}-\!-\!-\text{C}\sim \\
|\qquad|\qquad| \\
\text{H}\qquad\text{C}=\text{O}\;\;\text{H} \\
| \\
\text{O} \\
| \\
\text{H}-\text{C}-\text{H} \\
| \\
\text{H}
\end{array}
$$

The polymer is a colourless, transparent material of good toughness. Its optical properties are important and although it is softer and less wear resistant than glass, it often replaces glass for optical work.

The methyl CH_3 and ester $COOCH_3$ groups occur in a random fashion on each side of the molecule and, hence, the material is atactic and amorphous.

Perspex is soluble in solvents such as toluene, ethyl acetate, and trichlorethylene and, apart from its use in optics and as mouldings, it may be used dissolved in a solvent as a paint or varnish.

Some typical properties are given in Table 5.7.

Table 5.7

Property	High molecular weight type	Low molecular weight
Molecular weight	1 000 000	60 000
Density, kg/m³	$1{\cdot}19 \times 10^3$	$1{\cdot}18 \times 10^3$
Elastic modulus, MN/m²	2964·8	2413·2
Refractive index	1·49	1·49

Fluorocarbon polymers (linear crystalline thermoplastics)

These polymers combine high thermal stability with retention of flexibility down to very low temperatures. Many of them are chemically inert, have good electrical insulating properties and very low friction coefficients. The most representative member of the group is polytetrafluorethylene PTFE.

The monomer is the gas tetrafluorethylene

$$
\begin{array}{cc}
F & F \\
| & | \\
C & = C \\
| & | \\
F & F
\end{array}
$$

and this is polymerised under pressure using a free radical initiator to give PTFE

$$
\begin{array}{cccc}
F & F & F & F \\
| & | & | & | \\
\sim C - C - C - C \sim \\
| & | & | & | \\
F & F & F & F
\end{array}
$$

The orderliness of the chain molecule is at once apparent and, as would be expected, crystallinity is high. The molecular structure is very similar to that of polyethylene. It is the high regularity and crystallinity of PTFE that produces the remarkably high melting point of 600 K.

The bonds present, C–F and C–C, are both stable and so PTFE can operate for long periods of time at temperatures up to 520 K without change in properties.

The material is of high molecular weight, has up to 98% crystallinity and is practically insoluble in solvents. It is tough and will remain ductile at temperatures as low as 23 K.

The viscosity of PTFE at temperatures near or above its melting point is too high to allow it to be fabricated by the usual processes of extrusion or moulding. Most processing thus involves cold pressing of PTFE powder to shape, followed by sintering the pressing above the melting point to give a dense homogeneous shape. The technique is similar to that used in powder metallurgy.

Table 5.8

Density, kg/m^3	$2 \cdot 1$–$2 \cdot 3 \times 10^3$
Max. stress, MN/m^2	$17 \cdot 2$–$27 \cdot 6$
Elongation to fracture, %	200–300
Coefficient of friction	$0 \cdot 09$–$0 \cdot 12$

Typical properties of PTFE are given in Table 5.8. The polymer can be considerably strengthened by cold drawing to give orientation of the chain molecules. Its deformation resistance at elevated temperatures may also be improved by blending it with glass fibres during manufacture.

PTFE is used as an electrical insulating material, for chemical plant components and for non-lubricated bearings.

The high melting point and viscosity of PTFE requires a powder sintering technique for fabrication. To allow normal melt-processing techniques to be used, co-polymerisation with hexafluoropropylene

$$\begin{array}{ccc}
\text{F} & & \text{F} \\
| & & | \\
\text{C} & = & \text{C} \\
| & & | \\
\text{F} & & \text{F–C–F} \\
& & | \\
& & \text{F}
\end{array}$$

is sometimes used to give a material of reduced melting point and viscosity.

Polyamides—nylons (linear crystalline thermoplastics)

These differ from the polymers discussed previously in that they have atoms other than carbon in the molecule backbone. Basically, the molecules consist of the normal (CH_2) groups linked together with amide groups $(NH.CO)$.

The raw materials for the production of the nylons are:

(*a*) Adipic acid

$$HO-\overset{\overset{\textstyle O}{\|}}{C}-\overset{\overset{\textstyle H}{|}}{\underset{\underset{\textstyle H}{|}}{C}}-\overset{\overset{\textstyle H}{|}}{\underset{\underset{\textstyle H}{|}}{C}}-\overset{\overset{\textstyle H}{|}}{\underset{\underset{\textstyle H}{|}}{C}}-\overset{\overset{\textstyle H}{|}}{\underset{\underset{\textstyle H}{|}}{C}}-\overset{\overset{\textstyle O}{\|}}{C}-OH$$

(*b*) Hexamethylenediamine

$$N-\overset{\overset{\textstyle H}{|}}{\underset{\underset{\textstyle H}{|}}{C}}-\overset{\overset{\textstyle H}{|}}{\underset{\underset{\textstyle H}{|}}{C}}-\overset{\overset{\textstyle H}{|}}{\underset{\underset{\textstyle H}{|}}{C}}-\overset{\overset{\textstyle H}{|}}{\underset{\underset{\textstyle H}{|}}{C}}-\overset{\overset{\textstyle H}{|}}{\underset{\underset{\textstyle H}{|}}{C}}-\overset{\overset{\textstyle H}{|}}{\underset{\underset{\textstyle H}{|}}{C}}-N$$

(*c*) Sebacic acid

$$HO-\overset{\overset{\textstyle O}{\|}}{C}-\overset{\overset{\textstyle H}{|}}{\underset{\underset{\textstyle H}{|}}{C}}-\overset{\overset{\textstyle H}{|}}{\underset{\underset{\textstyle H}{|}}{C}}-\overset{\overset{\textstyle H}{|}}{\underset{\underset{\textstyle H}{|}}{C}}-\overset{\overset{\textstyle H}{|}}{\underset{\underset{\textstyle H}{|}}{C}}-\overset{\overset{\textstyle H}{|}}{\underset{\underset{\textstyle H}{|}}{C}}-\overset{\overset{\textstyle H}{|}}{\underset{\underset{\textstyle H}{|}}{C}}-\overset{\overset{\textstyle H}{|}}{\underset{\underset{\textstyle H}{|}}{C}}-\overset{\overset{\textstyle H}{|}}{\underset{\underset{\textstyle H}{|}}{C}}-\overset{\overset{\textstyle O}{\|}}{C}-OH$$

(*d*) Caprolactam

$$(CH_2)_5 \overset{\diagup CO}{\underset{\diagdown NH}{|}}$$

The three main molecular combinations of commercial importance are nylon 6, nylon 66 and nylon 610, the numbers referring to the way in which the polymer molecule is built up.

Nylon 6 or polycaprolactam has a molecule in which a carbon atom in the backbone is replaced by another type of atom after every sixth position

$$\sim(CH_2)_5 - NH - CO - (CH_2)_5 - NH - CO\sim$$

Nylon 66, a polymer of hexamethylenediamine and adipic acid, has a molecule in which a foreign atom is substituted after every sixth position, but the alternation of (CH_2) and (CO) groups differs from that in nylon 6.

$$\sim(CH_2)_4 - CO - NH - (CH_2)_6 - NH - CO - (CH_2)_4 - CO - NH - (CH_2)_6\sim$$

Nylon 610 is a polymer of hexamethylenediamine and sebacic

acid, and the molecule backbone is interrupted after every sixth
and tenth position, i.e.

$$\sim(CH_2)_8-CO-NH-(CH_2)_6-NH-CO-(CH_2)_8-CO-NH-(CH_2)_6\sim$$

The differences in properties exhibited by these three types of
nylon can be attributed to the differences in molecular structure
and the differences in bonding between the molecules. The attractive force between one chain molecule and its neighbour is due
to the affinity between the hydrogen atoms in the amino groups
and the oxygen atoms on the chains. The degree of this hydrogen
bonding will govern such properties as melting point and rigidity
and the variation in hydrogen bonding is indicated in Fig. 5.20
(nylon 6), Fig. 5.21 (nylon 66), and Fig. 5.22 (nylon 610).

It is obvious that, of the three types, nylon 66 has the greatest
degree of hydrogen bonding while nylon 610 has the least.

The amine sites (CO.NH) in the chains are open to chemical
attack and so frequency of occurrence will influence chemical prop-

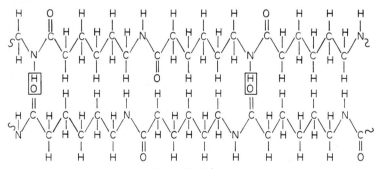

Fig. 5.20. Nylon 6

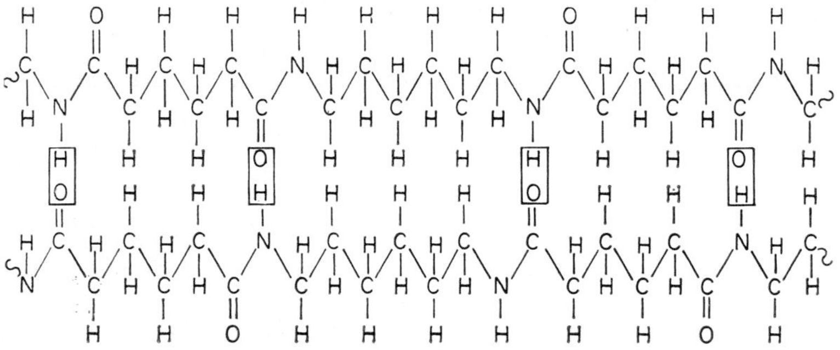

Fig. 5.21. Nylon 66

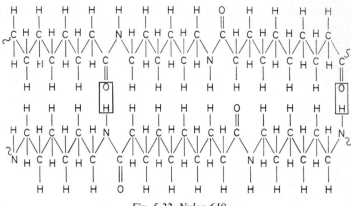

Fig. 5.22. Nylon 610

perties. For instance, nylon 610, which has the greatest spacing, absorbs least moisture and has the best chemical resistance. These amide groups also break up the exactness of matching between one chain and another and so influence the degree of crystallinity. One would therefore expect nylon 610 to be more crystalline than the other forms.

It is, of course, possible to produce co-polymers of these three main types. This usually produces more random arrangement of amide groups and less hydrogen bonding, and so such co-polymers are more flexible and more easily processed than the homopolymers.

Water absorption is a drawback with nylons since water acts as a plasticising agent and reduces both Tg and degree of crystalli-

Table 5.9

Property	Nylon 66	Nylon 6	Nylon 610	66/610/6 co-polymer 40%/30%/30%
Yield stress, MN/m²	79·3	75·8	58·6	31·0
% Elongation	90	160	120	300
Elastic modulus, MN/m²	2965	2758	2069	1379
Melting point, K	537	498	495	433
Density, kg/m³	1·14×10³	1·13×10³	1·09×10³	1·08×10³

nity. It is often found that on drying out, for example, nylons become much more rigid and undergo shrinkage.

Surface friction coefficients of nylons are about 0·4 when used against steel, as compared with 0·8 for steel–steel contacts. The nylons are therefore useful for bearings, the upper working loads being limited only by the build-up of frictional heat, resulting from the inherently low thermal conductivity of plastics. Some typical properties of nylons are given in Table 5.9.

The nylons are widely used in engineering for gears, cams, bearings and non-lubricated silent-moving parts. Nylon cord is used in tyre manufacture. The materials are fairly good electrical insulators but since the insulating properties deteriorate as moisture is absorbed, they are used as secondary insulation only.

Phenolic resins (cross-linking thermosetting polymers)

These are well-established moulding materials (bakelite) and are produced by condensation polymerisation reactions between phenol and formaldehyde.

Phenol is a solid obtained as a by-product from coal tar distillation or produced synthetically from benzene.

Formaldehyde is normally a gas but is soluble in water to give a solution known as formalin. It is produced by oxidation of methyl alcohol.

The condensation polymerisation process has been illustrated on page 173.

For commercial purposes, it may first be necessary to produce a comparatively low molecular weight polymer which can be blended with the necessary additives and then treated at a later stage to give the high molecular weight cross-linked material. These initial low molecular weight materials are either resols or novolaks.

The resol is prepared by reacting phenol with an excess of formaldehyde under alkaline conditions. It is a low polymer which will cross-link simply by heating. A probable molecular arrangement in a resol is shown in Fig. 5.23.

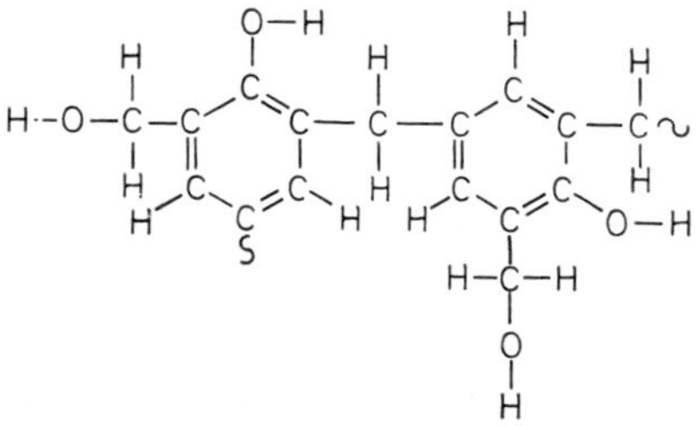

Fig. 5.23. A probable molecular arrangement in a resol

If resols are used, all the materials necessary to produce the commercial resin are placed in the reaction vessel and cross-linking is achieved simply by continued heating, i.e. one-stage resins.

The novolak is prepared by reacting formaldehyde with an excess of phenol under acid conditions. The low polymer does not cross-link on heating alone but needs to be heated in the presence of more formaldehyde. Fig. 5.24 indicates the molecular structure of a novolak.

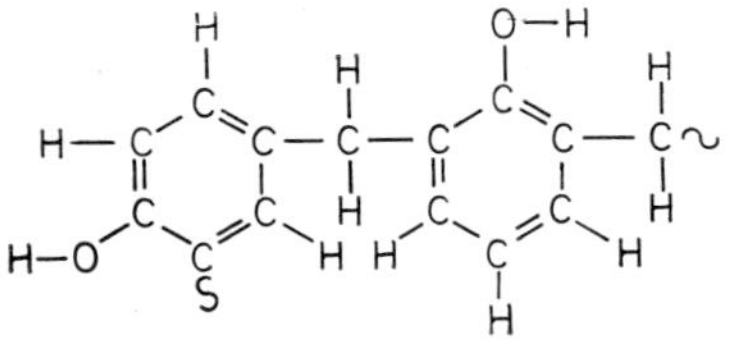

Fig. 5.24. The molecular structure of a novolak

The simple novolaks are mixed with the fillers and other necessary ingredients and the commercial resin produced by adding more formaldehyde and heating, i.e. two-stage resins.

The resins produced at the end of these processes are highly cross-linked but are still fusible under heat and pressure. The resins are therefore capable of being hot moulded into shape and this process also completes the cross-linking to give a final, infusible thermoset. The cross-linking in the resins occurs by bridging of (CH_2) groups between neighbouring benzene rings and, since

Table 5.10

Property	Normal grade	Electrical grade
Tensile strength, MN/m²	55·2	74·6
Density, kg/m³	$1·35 \times 10^3$	$1·85 \times 10^3$
Voltage to cause conduction across $2·5 \times 10^{-5}$ m at 293 K	150–300	275–3350

the bridging may occur at different points around the ring, a three-dimensional network molecule is produced. Some typical properties of phenolic resins are given in Table 5.10.

The polymers are hard and rather brittle and have good electrical insulating properties. They are widely used in moulded form for plugs, switches and as impregnating agents in the production of laminates, brake-linings, abrasive discs. The material is also used in the shell-moulding process of metal casting as a bond for the sand particles.

Inorganic polymers

Silicon, like carbon, is a tetravalent element from group IVA of the Periodic Table and is capable of co-valent bonding. Some polymers are thus available with silicon atoms forming the backbone of the chain molecules.

Compounds of the type

$$\sim \underset{\underset{H}{|}}{\overset{\overset{H}{|}}{Si}} - \underset{\underset{H}{|}}{\overset{\overset{H}{|}}{Si}} - \underset{\underset{H}{|}}{\overset{\overset{H}{|}}{Si}} - \underset{\underset{H}{|}}{\overset{\overset{H}{|}}{Si}} \sim$$

do exist but the groupings become unstable with above six silicon atoms and such a chain length could hardly be regarded as a polymer.

The siloxane grouping is, however, much more stable, i.e.

$$\sim \underset{\underset{R}{|}}{\overset{\overset{R}{|}}{Si}} - O - \underset{\underset{R}{|}}{\overset{\overset{R}{|}}{Si}} - O - \underset{\underset{R}{|}}{\overset{\overset{R}{|}}{Si}} - O \sim$$

15

merised structure. These polymerised ceramics may be either amorphous or crystalline depending on the degree of ordering.

Since the bonds which are present in ceramics do not involve the formation of free electrons, the materials have low thermal conductivity and are electrical insulators.

CRYSTALLINE CERAMIC PHASES

The crystal structures involved in ceramic materials, as mentioned earlier, are rather complex because of the presence of atoms or ions of different sizes. However, some of the more simple structures can be considered in illustration.

Periclase MgO

This phase forms the basis of the magnesite refractories used for basic furnace linings. The crystal structure consists of a face-centred cubic arrangement of the O^{2-} ions with Mg^{2+} ions in the interstitial positions. Fig. 6.1 gives a plan view of the pattern, the

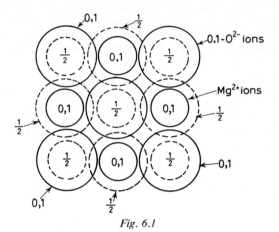

Fig. 6.1

numbers referring to the depth location of the ion within the unit cell. The same type of crystal structure exists in CaO which forms the basis of refractories containing lime.

In the unit cell of MgO shown there are:

(*a*) 8 corner O^{2-} ions, each shared by 8 unit cells, and so this particular cell actually contains $8 \times \frac{1}{8} = 1$ such oxygen ion.

(b) 6 facial O^{2-} ions, each shared by two unit cells, and so this particular cell controls $6 \times \frac{1}{2} = 3$ such facial ions.

The cell thus actually contains 4 oxygen ions.

(c) 12 facial interstitial Mg^{2+} ions, each shared by 4 unit cells, and so this particular cell contains $12 \times \frac{1}{4} = 3$ Mg^{2+} ions.

(d) 1 Mg^{2+} ion in the centre of the cell body.

The cell thus actually contains 4 Mg^{2+} ions. This sort of structure is sometimes referred to as the AX type, since it contains equal numbers of ions of each type.

Not all AX phases are ionically bonded. The refractory material silicon carbide, for example, is an AX-type compound but both Si and C have four co-valent bonds. The structure is then composed of carbon atoms in face-centred cubic positions with silicon atoms in those interstices which provide them with four carbon atoms as near neighbours.

Silica SiO_2

Silica forms the basis of many acid refractories such as sand, fireclay and firebrick. It is an allotropic material and can exist in three forms as quartz, crystobalite or tridymite.

The various forms are crystalline and contain Si^{4+} ions surrounded tetrahedrally by four O^{2-} ions. The unit cell of tridymite is shown in Fig. 6.2.

In materials based on silica, the SiO_2 is usually present as a silicate in which the actual arrangement of the silicon and oxygen

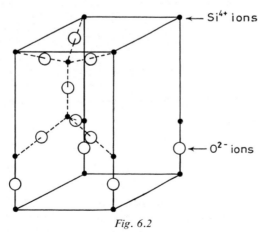

Fig. 6.2

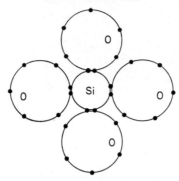

Fig. 6.3

is in the form of the (SiO_4) tetrahedron. This (SiO_4) group is co-valently bonded within itself, since electron sharing is involved, as shown in Fig. 6.3. This (SiO_4) group is capable of accepting four more electrons to achieve the stable octet arrangement and so is capable of becoming an SiO_4^{4-} ion. The electron deficiency in the (SiO_4) group may be satisfied by two mechanisms:

(a) The SiO_4 group can combine chemically with metal atoms. For example, if it combines with magnesium atoms, each metal atom gives up two electrons (since it is a divalent metal) and so becomes a magnesium ion Mg^{2+}. The four electrons made available convert the SiO_4 group into an SiO_4^{4-} ion and the net result is an ionically bonded silicate Mg_2SiO_4. This is a refractory material called forsterite.

(b) If insufficient metal ions are available, then polymerisation may occur. If the polymerisation is only partial, the $Si_2O_7^{6-}$ ion may be produced

$$-O-\underset{\underset{O}{|}}{\overset{\overset{O}{|}}{Si}}-O-\underset{\underset{O}{|}}{\overset{\overset{O}{|}}{Si}}-O-$$

This may then combine with metals to give complex metal silicates, e.g. $Ca_2MgSi_2O_7$, the three divalent metal ions supplying the necessary six electrons to give the ionic bond.

More complete polymerisation of the SiO_4^{4-} tetrahedron can occur producing long chains, as with carbon-based polymers. Often, these chains are cross-linked one to the other by ionic bonds provided by the presence of metal ions, as shown in Fig. 6.4.

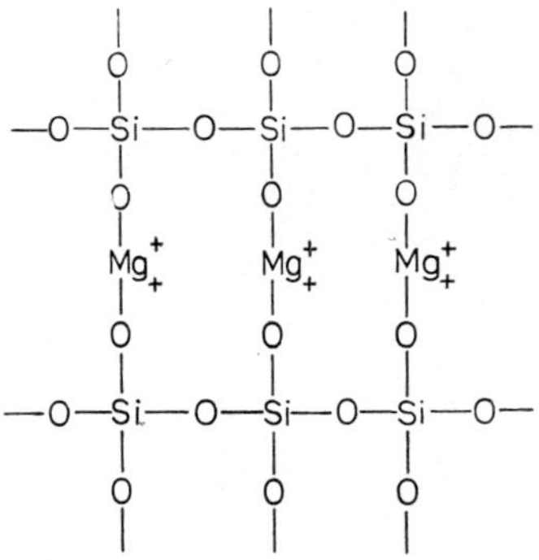

Fig. 6.4

The ionic bond across the chains is not as strong as the co-valent bonds along the chains and such structures tend to cleave easily parallel to the chains. The fibrous nature of asbestos can be attributed to this sort of internal structure.

Extension of the polymerisation of the SiO_4^{4-} group in two dimensions produces sheet-like structures typical of clay minerals, mica, talc, etc. Fig. 6.5 shows a plan view of one such sheet.

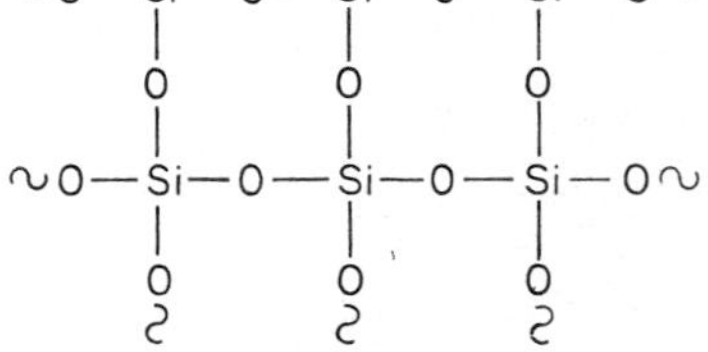

Fig. 6.5

If one extends this sheet *ad infinitum*, it is obvious that the oxygen atoms in between the silicon atoms have fully satisfied bonds. Thus, if two such sheets lie on top of each other, the only forces holding them together will be van der Waals' forces and so these sheets can easily slide over each other, particularly if they are held further apart than usual by the presence of foreign molecules. This sort of

structural arrangement is present in clay minerals and explains
the plasticity of clays when water molecules are added.

The picture given above is not truly representative since it must
be remembered that the Si atoms form the centres of tetrahedrons
with oxygen atoms at the apexes. The sheet is therefore three atoms
in depth.

One of the simplest of the clay minerals is kaolinite $Al_2Si_2O_5(OH)_4$.
This is a combination of the SiO_4^{4-} group and $Al(OH)_3$. $Al(OH)_3$
or Gibbsite also has a sheet-like structure and so kaolinite has
a complex layered structure which can be represented as:

$$6 \quad O^{2-} \text{ ions}$$

$$4 \quad Si^{4+} \text{ ions}$$

$$4 \quad O^{2-} + 2(OH)^- \text{ ions}$$

$$4 \quad Al^{3+} \text{ ions}$$

$$6 \quad (OH)^- \text{ ions}$$

Polymerisation of the SiO_4^{4-} tetrahedron in three dimensions is
also possible, giving rise to network structures co-valently bonded
in all dimensions. The structure of tridymite mentioned earlier is
really this sort of arrangement.

NON-CRYSTALLINE CERAMICS

An amorphous ceramic or glass is basically a supercooled liquid
of very high viscosity. There is, therefore, no long-range ordering
of the atoms.

Silica glass is the most common example of an amorphous ceramic based on a single oxide. Like crystalline silica, silica glass is
built up by repetition of the SiO_4^{4-} tetrahedron in three dimensions,
but unlike crystalline silica, there is no long-range order in the
network. This random network produced by three-dimensional
polymerisation of the SiO_4^{4-} group is shown in Fig. 6.6. Although
based on SiO_4^{4-}, the diagram shows 9 Si atoms and 18 O atoms
all fully satisfied as regards bonds, giving a chemical formula SiO_2.
All the bonds here are co-valent and so the material is likely to be
extremely hard and strong.

Most glasses contain metal oxides as well as silica and these
metal oxides tend to disrupt the network polymer arrangement of
pure SiO_2, since metal oxides are ionically bonded. Hence, the

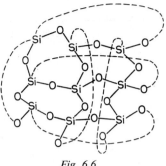

Fig. 6.6

addition of, say, sodium oxide to quartz glass would introduce Na^+ ions into the structure, resulting in the replacement of some of the co-valent bonds by ionic bonds. Fig. 6.7 indicates this.

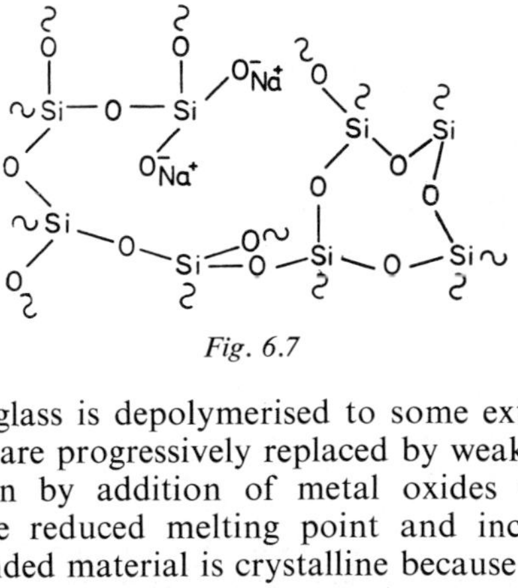

Fig. 6.7

In effect, the glass is depolymerised to some extent and strong co-valent bonds are progressively replaced by weaker ionic bonds. Depolymerisation by addition of metal oxides would thus be expected to give reduced melting point and increased fluidity. An ionically bonded material is crystalline because of the ordering necessary to maintain change balance. As more metal oxide is added to a glass, the ionic bonding increases at the expense of co-valent bonding and such glasses will crystallise more readily. The slags used in metal refining and welding processes are typical examples of glasses heavily loaded with metal oxides. The compositions of some typical glasses are given in Table 6.1.

The supercooled liquid condition which exists in highly polymerised glasses is a metastable condition and, theoretically, any glass could be made to revert to its stable, crystalline condition by very slow cooling from the liquid condition or by prolonged annealing. However, crystallisation rate varies exponentially with

Table 6.1

Type	SiO$_2$	Al$_2$O$_3$	CaO	Na$_2$O	B$_2$O$_3$	MgO	Others	
Optical glass	50–70			1–8	10		BaO	2–3
							K$_2$O	8–10
							ZnO	8
Pyrex glass	82	2			4	12		
Quartz glass	99							

temperature according to an Arrhenius-type relationship and is usually negligible. Crystallisation or devitrification does, however, become easier as the degree of polymerisation is reduced. It is usually to be avoided since it involves loss of transparency and may cause internal fractures.

CERAMIC PHASE DIAGRAMS AND THEIR USE

In the production of ceramic bodies, such as refractories for furnace linings, the component minerals are first mixed together and then pressed to give the required shape. To develop the high strength needed in such a body it is then necessary to fire the material at high temperature. This often involves changes of phase and partial fusion. The liquid phase so produced may then, on freezing, act as a bond between the other parts of the ceramic body. In this respect, therefore, phase diagrams are useful in predicting changes in constitution.

Some typical examples of the use of such diagrams are given below.

The SiO$_2$–Al$_2$O$_3$ system

Many furnaces are lined with acid refractories, i.e. refractories which resist attack by acid (siliceous) slags at high temperatures. These refractories are often based on silica–alumina mixtures and so Fig. 6.8 should give some information on the properties of such alumino–silicate refractories.

The diagram indicates a eutectic reaction at 2113 K which produces mullite and corundum Al$_2$O$_3$. A further eutectic reaction at 1868 K produces a mullite–cristobalite mixture. The phase mullite is a solid solution and has a composition near to Al$_6$Si$_2$O$_{13}$.

Cristobalite is the high temperature allotrope of SiO_2 and at equilibrium rates of cooling, transforms to tridymite below 1743 K.

Normal fireclay contains about 60% SiO_2, 40% Al_2O_3, as indicated by line *A–A* in Fig. 6.8. On slowly raising the temperature of such a body, an allotropic change is indicated at 1743 K, but

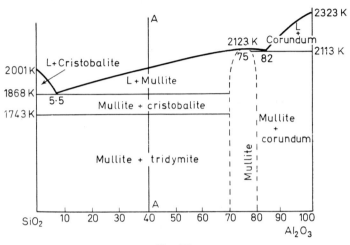

Fig. 6.8

this may not occur since the change is very sluggish. At 1868 K the eutectic is reached and use of the lever law indicates that nearly half the material liquefies to give eutectic containing 5·5% Al_2O_3. This obviously places a restriction on the upper working temperature of such a refractory. This temperature is also that at which the refractory would be fired after pressing to shape and so its structure will consist of grains of crystalline mullite bonded with eutectic. The eutectic should theoretically also be crystalline but crystallisation is so slow that it is usually present as a metastable glass.

Refractories within the composition range 0–72% Al_2O_3 will always produce eutectic liquid at 1868 K but the proportion of eutectic will vary. The eutectic glass in such materials will deform under load at sub-eutectic temperatures and this is aggravated if metal oxides are present, since these lower the melting range still further. Absorption of metal oxides can, of course, quite easily occur in service.

Refractories in this system which are required to have high temperature stability and load-bearing properties would need to contain over about 72% Al_2O_3. Liquid is not produced in these refractories (mullite and high alumina) until about 2113 K. The liquid eutectic in these materials crystallises rapidly, since it contains so little SiO_2 and so is not present as a glass. This reduces the possibility of viscous flow occurring at sub-eutectic temperatures.

The SiO_2–CaO system

Fig. 6.9 shows part of this system.

Silica refractories are widely used for acid linings. About 2% of lime CaO is added to produce a little $CaSiO_3$ glass to act as a bond.

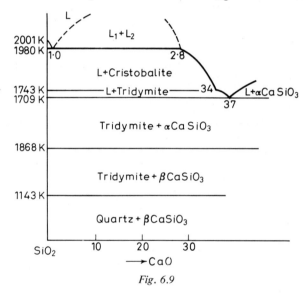

Fig. 6.9

One disadvantage of SiO_2 is its polymorphism. It can exist in three main forms, i.e. quartz, tridymite and cristobalite. Theoretically, quartz exists up to 1143 K, tridymite from 1143 to 1743 K and cristobalite from 1743 K to the melting point. In fact, the transformations are so sluggish that on heating, for example, quartz may change directly to cristobalite without producing tridymite and, at room temperature, the material may contain all three allotropes existing together. These three main allotropes of SiO_2 are the result of long-range rearrangement of the Si and O atoms, i.e. they are reconstructive transformations. Besides these, there are other minor changes |which can occur within each allotrope. These changes involve only slight displacements of the Si and O atoms and so occur fairly rapidly. For example,

α quartz exists below 846 K $\left.\right\}$ 1·4% volume change
β quartz exists above 846 K $\left.\right\}$

α tridymite exists up to 390 K

β_1 tridymite exists between 390 and 436 K

β_2 tridymite exists above 436 K

Cristobalite may exist as α or β, the change taking place over the range 473 to 548 K and producing a 3·5% volume change.

Both the main allotropic changes and the sub-changes involve changes in density and, hence, volume changes. The transformation of quartz to cristobalite, for instance, involves a 15% expansion and so it is obvious that quartz should not be present in the finished refractory, otherwise the resultant volume changes on heating could easily cause spalling and cracking. The preferred form of SiO_2 in the final refractory is tridymite. This is metastable up to 1143 K and stable from 1143 to 1743 K and has the lowest $\alpha - \beta$ volume change of all the allotropes.

The mixture for silica refractory manufacture consists of natural quartz with a little lime. The lime provides the $CaSiO_3$ bond and at the same time catalyses the change of quartz into tridymite on firing. The final refractory will, however, usually consist of a mixture of cristobalite and tridymite.

The SiO_2–CaO diagram indicates that the firing temperature should be about 1743 K in order to develop liquids. The slope of the liquidus curve from 1743 to 1980 K is high and use of the lever law here would indicate that continued heating well above 1743 K produces very little increase in the volume of liquid produced. It is, therefore, possible to operate silica refractories above 1743 K without danger of collapse due to liquefaction.

The chief danger with silica refractories is obviously that of spalling and cracking due to volume changes resulting from the allotropy of silica. Such refractories cannot, therefore, be used in conditions of rapidly fluctuating temperature.

The MgO–SiO₂ system

A basic refractory is one which will resist attack by slags containing high proportions of lime CaO or magnesia MgO. Most of these refractories are based on magnesia or periclase MgO. For example,

> Magnesite refractories—mainly MgO
>
> Dolomite refractories—CaO + MgO
>
> Forsterite refractories—MgO + SiO₂

Refractories based on MgO usually contain a little Al_2O_3 or SiO_2 to produce the bond and so consist of crystals of periclase MgO with forsterite Mg_2SiO_4 glass. Very little bond is actually present because, as Fig. 6.10 indicates, a temperature of about 2123 K would be needed to develop this. As a result, magnesite refractories may be quite porous.

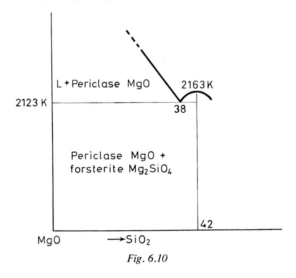

Fig. 6.10

The Na₂O–SiO₂ system

The depolymerisation of glasses by metal oxides has already been discussed. The effect is a progressive replacement of co-valent bonds by ionic bonds and this results in a lowering of fusion temperature and viscosity. Fig. 6.11 indicates that this behaviour is the result of eutectic formation.

The addition of Na_2O to silica, as in optical glass manufacture, will therefore lower the softening point from the melting point of SiO_2 at 1996 K to about 1053 K, the fusion point of the quartz–sodium silicate eutectic. The proportion of eutectic will increase as the Na_2O content increases and hence fluidity will increase.

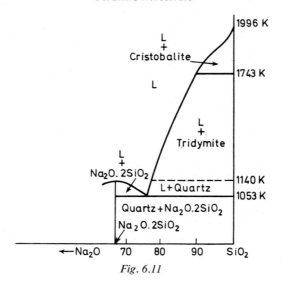

Fig. 6.11

The SiO_2–FeO system

A similar sort of situation arises in the SiO_2–FeO system, as indicated in Fig. 6.12. This system represents the behaviour of acid slags in steelmaking. The basis of such a slag is sand, but during

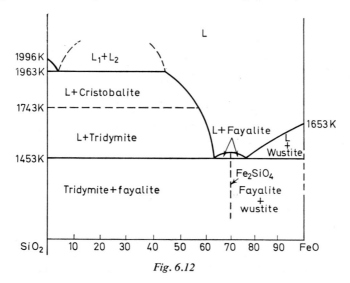

Fig. 6.12

the refining process iron oxide is picked up from the melt or is deliberately added during the refining stage.

The phase Fe_2SiO_4 is fayalite, while wustite is FeO. The addition of FeO to silica lowers the fusion temperature from 1996 K to below 1470 K. The diagram would therefore predict a loss of refractoriness of sands if attacked by metal oxide. It also indicates that the fluidity of an acid slag is a function of its FeO content. Generally, a fluid slag is more active as a refining agent than is a thick, viscous slag.

THE RESPONSE OF CERAMIC MATERIALS TO STRESS

Many ceramic phases are crystalline in nature and will therefore contain dislocations, as do metal crystals. However, unlike metals, ceramics show almost no capacity to undergo slip and plastic deformation. As a result, they are usually extremely brittle materials. The reason for this difference between metals and ceramics lies in the fact that whereas the atoms in metal crystals are equivalent, a ceramic crystal is usually built up from ions of unlike charge.

Fig. 6.13 illustrates the cubic structure of periclase MgO containing an edge dislocation which disrupts the regular alternation of Mg^{2+} and O^{2-} ions in the lattice. It is obvious that any attempt to slip in the [100] direction will be strongly resisted since this would mean bringing ions of like charge into closer contact.

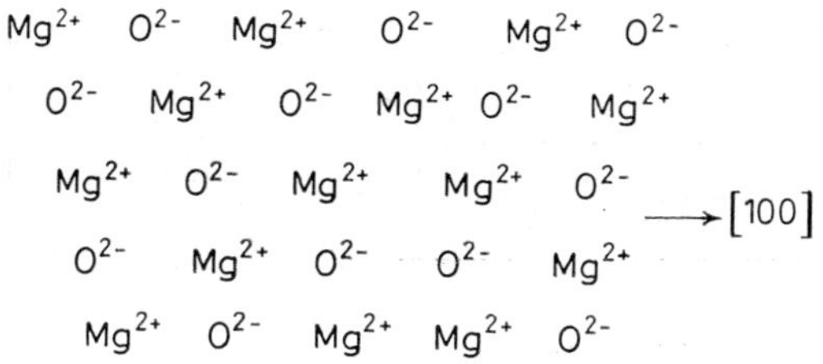

$$Mg^{2+} \quad O^{2-} \quad Mg^{2+} \quad O^{2-} \quad Mg^{2+} \quad O^{2-}$$
$$O^{2-} \quad Mg^{2+} \quad O^{2-} \quad Mg^{2+} \quad O^{2-} \quad Mg^{2+}$$
$$Mg^{2+} \quad O^{2-} \quad Mg^{2+} \quad Mg^{2+} \quad O^{2-} \quad \longrightarrow [100]$$
$$O^{2-} \quad Mg^{2+} \quad O^{2-} \quad O^{2-} \quad Mg^{2+}$$
$$Mg^{2+} \quad O^{2-} \quad Mg^{2+} \quad Mg^{2+} \quad O^{2-}$$

Fig. 6.13

There is, as a result, a severe restriction on the availability of slip systems in ionically bonded materials and such materials will tend to fracture at lower stresses than those at which slip would occur.

Since slip is restricted, and since very strong co-valent and ionic bonds are present, one might expect that ceramic materials would have very high strengths and elastic moduli. The moduli are, in fact, higher than in other materials and the compressive strength also reflects the bond strength. The tensile strength is, however, remarkably low. This state of affairs is typical of brittle materials and is usually explained by using the Griffith concept of the stress-concentrating effect of inherent surface flaws. Griffith showed that the inherent faults or microcracks in a brittle material could concentrate the applied stress such that the theoretical strength is locally surpassed causing a crack which would then spread. He was able to calculate the value of the concentrated stress σ_c at the root of the fault as

$$\sigma_c = 2\sqrt{\frac{L}{R}} \cdot \sigma_A$$

where L is the length of the inherent flaw; R is the root radius of the flaw, and σ_A is the applied stress normal to the plane of fracture.

The inherent flaws in a brittle material are very tiny with root radii of atomic dimensions—say 2×10^{-10} m. Such faults can obviously produce a dramatic concentration of the applied stress and so the crack will open out and spread very rapidly. A running crack could be stopped by blunting of its root by plastic deformation but in a brittle material the ratio L/R will continue to increase as the crack spreads.

In compression, of course, any inherent faults will tend to be closed up and so the material may show very high strength.

Non-crystalline ceramics, such as glasses, also fracture in tension because of this same stress-concentrating effect of surface flaws.

According to the Griffith concept, the less surface area there is present, the stronger the material should be, since there is then less chance of flaws occurring. This is found to be true and very fine filaments of glasses and ceramics have remarkably high tensile strengths.

Since amorphous ceramics such as glasses do not contain dislocations, they will not undergo plastic deformation by the slip process but by a process of viscous flow. This occurs easily at high temperatures. Viscous flow in glasses can be visualised as a progressive breaking and rearrangement of the bonds between the atoms. It has previously been indicated that the structure of solid glass consists of a random network of atoms showing only short-range order (Fig. 6.6). Interatomic distances, therefore, vary widely and so some bonds will be inherently more highly stressed than others. Application of external stress may rupture these bonds and then the neighbouring atoms must rearrange their spacing. Such rearrangement involves extension of more bonds which then rupture in their turn. A steadily applied load will therefore cause continuous deformation by viscous flow, particularly if the glass is at elevated temperature or if it has been depolymerised by addition of metal oxides.

Since viscous flow deformation of glass involves the movement and rearrangement of atoms, then

(*a*) The rate of flow will be proportional to the applied stress, i.e.

$$\mu = \frac{\sigma}{\dfrac{\mathrm{d}e}{\mathrm{d}t}}$$

where μ is the coefficient of viscosity; σ is the applied shear stress; $\mathrm{d}e/\mathrm{d}t$ is the rate of change of deformation with time.

(*b*) The viscosity will be inversely proportional to temperature via an Arrhenius relationship, i.e.

$$\mu = A\mathrm{e}^{E/kT}$$

(*c*) The response to stress will depend on the rate of loading. Thus shock loading a glass, even at elevated temperature, will still cause it to fail in a brittle manner, because time for bond rearrangement has not been given.

The stresses on many ceramic materials, particularly refractory materials, are not mechanically applied but are thermal stresses resulting from inhibited expansion or contraction. Thermal properties are therefore particularly important. The non-uniform dimensional changes induced by temperature are more significant in ceramics than in other materials simply because ceramics lack the inherent ductility to absorb and redistribute these stresses. Such non-uniform changes in dimensions may be the result of one or more of the following factors.

(*a*) *Phase transformation*—This is typical of the changes in SiO_2, where both the major reconstructive changes and the minor displacive changes occur. The different allotropes of SiO_2 have different densities and, hence, volume changes are involved. The quartz to cristobalite change, for example, involves a 15% volume expansion. Obviously, rapid temperature changes in such materials must be avoided.

(*b*) *Anisotropy of expansion characteristics*—In hexagonal crystal structures such as exist in corundum Al_2O_3, different crystallographic directions will have different coefficients of expansion as a result of different bond strengths. The expansion is greater perpendicular to the close-packed planes than parallel to them. This sort of anisotropy can, of course, be evened out by using a finely crystalline material.

(c) *Differential expansion coefficients*—Ceramic bodies are often made up from different phases, e.g. crystals of one phase embedded in a glassy matrix. These different phases will have different expansion coefficients and sufficient thermal stress may be generated to cause internal cracking if the material is heated or cooled rapidly.

(d) *Low thermal conductivity*—Ceramic phases have thermal conductivities which are 10 to 100 times smaller than those found in metals. There is then always the possibility that severe temperature gradients may be developed in ceramics on rapid heating or cooling and the resultant differential expansion characteristics can cause spalling or flaking. There is also the problem that the different phases present in a ceramic body will have different thermal conductivities.

BIBLIOGRAPHY

DAVIES, T. H., *The Science of Engineering Materials (The Physics of Glass)*, Wiley, New York (1957)

DOUGLAS, R. W., *Properties and Structure of Glass* (Progress in Ceramic Science, Vol. 1), Pergamon, New York (1961)

'Engineering properties of ceramic materials' (Ed. Lynch), American Ceramic Society (1966)

KINGERY, W. D., *Introduction to Ceramics*, Wiley, New York

'Metals, plastics, ceramics—competition or complementation', *Metals and Materials*, Metals and Metallurgy Trust, London (Sept. and Oct. 1968)

NORTON, F. H., *Elements of Ceramics*, Addison-Wesley, Massachusetts (1957)

Science of Ceramics (Ed. Stewart, G. H.), Vols 1–3, Academic Press (1962; 1965; 1967)

Scientific American, Vol. 217 (Sept. 1967)

VAN VLACK, L. H., *Physical Ceramics for Engineers*, Addison-Wesley (1964)

Chapter 7

COMPOSITE MATERIALS

Composite materials are engineering materials which aim to combine, into a single body, the desirable properties of a number of separate materials.

In the broadest sense, practically all materials are composite to some extent. For example, engineering plastics usually contain additives such as fillers, and metallic alloys usually contain more than one phase. There are also the more easily recognisable composites, such as clad metals for increased corrosion resistance, hard-faced alloys for wear resistance and the honeycomb–resin bonded sandwiches now being used in aircraft construction.

It is obviously not possible in a single chapter to cover all forms of composite materials and so the discussion has been limited to those materials which consist of intimate three-dimensional mixtures. This means that surface-modified materials, such as carburised steels, plastic-coated metals, plated materials and so on, are not included.

The materials which are covered in this chapter are:

(*a*) Dispersion-strengthened materials, consisting of a matrix containing up to about 15% of particles of a different material, the particles being up to 0·1 μ in size.
(*b*) Particle-reinforced materials, consisting of a matrix containing over 20% of particles of size greater than 1·0 μ.
(*c*) Fibre-reinforced materials, consisting of a matrix with anything up to 70% by volume of fibres of a different material embedded in it.

Many of these composite materials are manufactured by a powder-sintering technique. In this technique, finely divided powders of different materials are mixed together, compacted by pressing in a die, and finally consolidated by heating (sintering).

The temperatures used are usually well below the melting point of the powders, since the sintering operation relies upon recrystallisation of the cold-worked powder particles, such recrystallisation and grain growth bridging the interface between one powder particle and its neighbours.

DISPERSION-STRENGTHENED COMPOSITE MATERIALS

These are materials composed of a metal matrix in which are embedded fine particles of oxides, carbides, nitrides or borides which are insoluble in and incoherent with the matrix.

The matrix is the main load-bearing constituent and the strengthening effect of the particles arises from their ability to prevent dislocation motion through the lattice, so producing higher room temperature strength and higher elevated temperature strength.

In order that the particles shall provide resistance to motion of dislocations through the matrix, they must obviously be of small enough size and be finely enough dispersed. Since it is also, in many cases, desirable that the properties of the matrix material shall be preserved, the volume fraction of particles must be limited. To fulfil these conditions, particle diameters are usually below $0 \cdot 3$ μ, with volume fractions below 15%.

It can be shown from dislocation theory that the smallest radius of curvature R to which a dislocation loop can be bent under an applied shear stress τ is

$$R = \frac{Gb}{2\tau}$$

where G is the shear modulus of the lattice, and b is the Burgers vector of the dislocation.

Thus, if the particle spacing D is taken as $2R$, then $\tau = (G.b)/D$ and this is the shear stress required to bow the dislocation line between the particles, as indicated in Fig. 7.1(a).

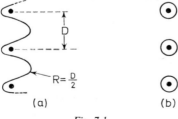

(a) (b)

Fig. 7.1

The whole dislocation line cannot pass through and so the original line breaks up, leaving a ring of slipped lattice around each particle, before passing on, as shown in Fig. 7.1(b). This, in effect, reduces the inter-particle spacing still further and any other dislocation line approaching the particles will need an even higher shear stress to force it through. The net result is that the material exhibits a very high rate of work hardening.

If it is assumed that the particles are so widely spaced that a dislocation line meets no resistance, then the stress needed to move the line would simply be the yield stress of the matrix. For metals, this is typically about $G/1000$ and hence $\tau = G/500$. The Burgers vector b is typically about 3×10^{-10} m, hence

$$D = 2R = 2\left[\frac{Gb}{2\tau}\right] = 2\left[\frac{G \times 3 \times 10^{-10}}{2\left(\dfrac{G}{500}\right)}\right]$$

$$= 0 \cdot 15 \times 10^{-6} \text{ m} = 0 \cdot 15 \ \mu$$

The inter-particle spacing must obviously be below this value in order that the particles shall interfere with dislocation motion.

Since the particles are insoluble in the matrix, transport of them by diffusion through the matrix is negligible and so they will not agglomerate on heating. This means that, unlike precipitation-hardened alloys, such materials will retain their strength at elevated temperatures and so may be useful as creep-resistant materials.

As regards commercial materials, the metal–metal oxides systems are the ones of major interest. A typical material is SAP, an aluminium–aluminium oxide composite. This is produced by surface oxidation of flake aluminium powder particles about $0 \cdot 1 \ \mu$ thick so that they become coated with an oxide film about 100×10^{-10} m thick. The oxidised flakes are then compacted by normal powder-sintering techniques. Such a material retains its usefulness and stability to temperatures very near the melting point of aluminium and is used in jet engine after-burners operating at up to 400°C.

Internal oxidation techniques are also available in which oxygen is diffused into a solid solution alloy so that the solute is preferentially oxidised, giving a very fine dispersion of oxide throughout the metal matrix. A Cu–Al alloy treated in this way to give about 4% by volume of Al_2O_3 particles gives better strength above 850 K than monel alloy and better strength above 950 K than even stainless steel.

It is also possible to produce a dispersed oxide in a metal matrix by mixing together the matrix metal oxide with the dispersed oxide and then selectively reducing the matrix oxide to metal just before powder sintering. This process is applied to the production of a nickel matrix with dispersed thorium oxide. Such a material has better high-temperature strength than the conventional nickel superalloys and has been used for the production of uncoated parts for jet engines working at temperatures up to 1420 K—temperatures not very far below the melting point of pure nickel.

These increases in high-temperature strength can be made even more pronounced if alloy matrices are used instead of pure metal.

PARTICLE-REINFORCED MATERIALS

These materials also consist of a matrix containing particles of another material but, in this case, both matrix and particles share the load. The particle size exceeds 1·0 μ with volume fractions above 25%. The particles are therefore too widely spaced to provide interference with dislocation motion in the matrix but produce a strengthening effect by mechanically restraining the matrix deformation. Particulate composites can be produced with either a metal matrix or with a polymer matrix.

The degree of restraint on matrix deformation produced by the particles is not accurately predictable but it is a function of the elastic properties of the matrix and particle. Ideally, the average modulus E_c of such a composite should be

$$E_c = V_m E_m + V_p E_p$$

where V is the volume fraction of matrix or particle. In actual fact, the modulus E_c is always less than this ideal. However, the composite is often used for purposes other than high strength and so this is not always of primary importance.

METAL-CERAMIC MIXTURES—CERMETS

These are materials which aim to blend the high strength of ceramics with the ductility and toughness of metals. They were originally developed to meet the demands of the nuclear-power and high-temperature engine industries for creep-resistant components. However, cermets have not yet been able to meet these demands because of their brittleness but they are widely used in applications where high strength and wear resistance are important, e.g. cutting tool tips, drilling bits, spark machining electrodes.

Cermets are usually produced by infiltrating liquid metal in some way around particles of ceramic. This means that a strong bond needs to be produced between the metal and the ceramic and ideally the bond should be a chemical bond produced either (*a*) by surface reaction between metal and ceramic to produce a bridging intermediate phase, or (*b*) by solid solution formation between metal and ceramic.

In order to achieve this bonding, it is necessary for the liquid metal to wet the surface of the ceramic particle. This wetting ability of a liquid is a function of the surface tension of the liquid and the surface energy of the solid and can be measured by the contact angle θ, as indicated in Fig. 7.2.

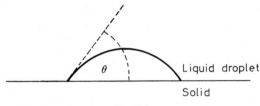

Fig. 7.2

In the ideal case, the contact angle would be zero but this is never achieved in most metal–ceramic mixtures. The surface nature of the ceramic particles is important here since films of foreign matter on them will rapidly increase θ.

The theoretical strength of the final composite is ideally the work needed to separate metal from ceramic. This value is found to be $W = \gamma L(1 - \cos \theta)$, where γL is the surface energy (tension) of the liquid metal originally used in making the composite. Even in mixtures where wetting ability was initially low, i.e. high values of θ, the relationship predicts final strengths of the order of 2700 MN/m². Complete wetting, giving very low values of θ, would indicate strengths up to 20 000 MN/m² but such strengths are never achieved even under ideal conditions, mainly because of the ease with which cracks can be generated from stress concentrations set up at the particle–metal interfaces. This is particularly the case if angular ceramic particles are used, and this situation is aggravated if there is a wide difference in the coefficients of expansion of the two phases. A difference of 5×10^{-6}°C⁻¹ can be tolerated, but anything greater than this produces thermal strains capable of causing internal fracture.

In general, therefore, cermets will have good high-temperature strength and creep resistance but will be rather sensitive to shock loading and to thermal shock. This inherent brittleness has so far prevented their application in the moving parts of high-temperature engines.

Typical types of cermet include:

(*a*) *Carbide types*—These are fabricated by standard powder-sintering methods and are used chiefly as cutting tools, e.g.

Ceramic	*Metal*
Silicon carbide SiC	Ag, Co, Cr
Titanium carbide TiC	Mo, W, Fe, Ni, Co
Tungsten carbide WC	Co

(*b*) *Oxide types*—These have very good high-temperature properties. For example, the $Cr–Al_2O_3$ composite has outstanding oxidation and wear resistance and is used extensively in sparking plug assemblies, as furnace linings, in rocket motor exhausts.

Ceramic	*Metal*
Aluminium oxide Al_2O_3	Al, Co, Fe, Cr
Chromium oxide Cr_2O_3	Cr
Magnesium oxide MgO	Al, Co, Fe, Mg
Silicon oxide SiO_2	Cr, Si

(*c*) *Boride types*—Borides have extreme hardness but poor oxidation resistance. Boride cermets are sometimes used as cutting tool tips.

Ceramic	*Metal*
Chromium boride Cr_3B_2	Ni
Titanium boride TiB_2	Fe, Ni, Co

As pointed out earlier, it was originally hoped that cermets would be usable for turbine blade manufacture in jet engines. Such engines

Table 7.1

Property	$77Cr : 23Al_2O_3$	$54TiC : 46Ni$	*Nimonic 105*
Density $\times 10^3$ kg/m^3	5·9	6·1	8·0
Expansion coefficient, $°C^{-1}$	8·9		19·7
Modulus of rupture at 293 K, MN/m^2	310	950	1030
Impact value, J	1·6	9·1	156

are more efficient the higher the gas temperature, and turbine blade material must possess a combination of high hot strength, good creep resistance, resistance to thermal shock and corrosion. Cermets cannot meet all these requirements and so cannot compete with superalloys such as nimonic. Table 7.1 compares some of the properties of cermets with a nimonic alloy.

Although the inherent brittleness prevents application in high-temperature engines, cermets are widely used as friction materials for brakes, as cutting tool tips and as fuel elements in nuclear power plant (uranium oxide–metal mixtures).

PARTICLE-REINFORCED POLYMERS

These are often referred to as filled polymers, the fillers being finely powdered carbon black, clays, silica, mica, etc. These parti-cles are added to improve such properties as surface hardness, shrinkage on moulding, ease of moulding and colour. They also, of course, reduce overall cost. These mineral fillers are widely used to stiffen otherwise rubbery polymers and so increase their useful stress range and temperature range of operation. Typical of this application is the addition of carbon black to natural and synthetic rubbers. The addition of 15–20% by volume of carbon gives a material about ten times as strong as the raw rubber and with a higher elastic modulus and abrasion resistance.

The opposite treatment is also possible, i.e. particles of rubber can be incorporated into otherwise rigid polymers to increase their toughness, particularly for low-temperature applications.

FIBRE-REINFORCED MATERIALS

These materials are made by embedding either continuous or discontinuous fibres of very hard material into a softer, more ductile matrix. The fibres are aligned so that they coincide with the direction of stressing of the finished component.

Conventional methods of strengthening usually rely on the inhibition of dislocation motion, either by mutual jamming (work hardening) or by the use of obstacles (age hardening, dispersion hardening).

Strengthening by incorporation of fibres differs from this, since the fibres are not primarily obstacles to dislocation movement. Under an applied stress, the plastic flow in the soft matrix throws load on to the fibres and, in fact, the fibres then carry the majority of the applied load. If high-strength high-modulus fibres are used, therefore, it should be possible to obtain very high strengths within the composite. When such a composite is tensioned parallel to the fibres, the principle of combined action comes into play, i.e. the strains in the matrix and in the fibres are almost equal. However, the stress–strain curves of the fibres and the matrix are very different from each other and, for any given common strain, the fibres will be carrying a much higher stress than the matrix. This difference is so marked that the contribution of the matrix can be ignored.

The matrix material, in this case, can be thought of as a load transmitter. In order to transmit load efficiently, the matrix must plastically deform at much lower stresses than the working stress of the composite and the fibres should be present as filaments or plates with high length/thickness ratios.

The matrix material must also bind the fibres together and protect them from surface damage and there must be sufficient separation between fibres so that any cracks within them are blunted and brought to a stop by plastic flow within the matrix. Typical fibres include the following.

Whiskers

A perfect lattice free from dislocations would have a strength about 1000 times the observed strength. Such a body would be produced in sections of crystalline materials only a single atom in thickness. As thickness increases, so does the possibility of dislocation content. It is, however, possible to produce whiskers of crystalline materials with diameters of the order of 1×10^{-7} m and these have remarkably high strengths simply because they contain fewer dislocations than usual. They are, of course, extremely brittle. Table 7.2 indicates the properties of typical whiskers.

Table 7.2

Whisker material	UTS MN/m^2	E MN/m^2	Melting point K
Graphite	20 600	675 710	3273
Al_2O_3	15 150	524 000	2323
Fe	12 430	193 000	1808
SiC	20 600	689 500	2873

Whisker materials are not widely used in composites. As one would imagine, they are extremely expensive, difficult to produce and align, and cannot be produced in anything but the smallest continuous lengths.

Ceramic and polymer fibres

Glass and ceramic fibres are very sensitive to surface condition and, as indicated in the section on ceramics, page 235, even the most minute surface flaw can initiate a Griffith crack in the presence of a tensile stress. Surface protection during fibre manufacture can be given by flame polishing after drawing down to the correct diameter or by coating each fibre with a vapour-deposited film of ductile metal after drawing. The fibres are produced by pulling or drawing them from the bulk material while it is hot enough to be in a viscous condition. Such fibres show an exponential decrease in viscosity and stiffness as temperature increases and so are limited to reinforcement procedures not involving temperatures above about 600 K.

Some crystalline polymer fibres have good room temperature strength which can be raised even further by cold drawing. They are not of much importance as reinforcement material because of the inherently low modulus and the rapid loss of strength as temperature increases.

Metal wire

Drawn or extruded metal filaments constitute the major source of reinforcement fibres. Extremely fine wires down to 1×10^{-6} m can be produced by sheathing bulk metal in glass, hot drawing

Table 7.3

Material	*UTS* MN/m²	*E* MN/m²	*Melting point* K
High carbon steel	3965	206 850	
Stainless steel	2413	200 000	
Titanium alloy	2206	117 215	
Tungsten	2894	344 750	3653
Aluminium	166·5	68 950	933
Copper	414	124 110	1356

and then removing the glass by etching. Even conventional cold-drawing techniques are capable of producing continuous filaments in diameters of the order of 1.5×10^{-4} m. Typical properties of cold-drawn wires of this diameter are given in Table 7.3.

Cold-drawn metal fibres are, of course, highly work hardened and would suffer a drastic loss of strength if heated to temperatures sufficient to cause recrystallisation.

Response of composites to stress

The fibres in a fibre-reinforced composite are usually aligned such that the service stress is applied parallel to them. It should be realised that any fibre-reinforced material will exhibit pronounced anisotropy as regards mechanical properties. The composite tends to be less strong in shear or compression than it is in tension. A compromise can be reached here by building-up the composite of laminations in which the fibres have different orientations. The response of the composites to applied stress will differ depending on the continuity of the fibres.

Continuous fibre reinforcement

If a progressively increasing tensile stress is applied, the response can be divided into four distinct stages:

(1) Elastic deformation occurs in both fibres and matrix.
(2) When the strain reaches the yield strain of the matrix material, the matrix begins to deform plastically but the fibres continue to deform elastically.
(3) Both matrix and fibres deform plastically.
(4) Fracture occurs in the fibres, the cracks spread very rapidly across the matrix and the composite fails.

In the elastic range of stage (1), the elastic modulus of the composite E_c is given by the usual rule of mixtures

$$E_c = E_f V_f + E_m V_m$$

where V refers to volume fractions, and the subscripts f, m refer to fibres and matrix, respectively. Increasing V_f will obviously give a higher composite strength.

In service, a composite will normally operate under stage (2) conditions and the modulus E_c during this stage is given by

$$E_c = E_f V_f + \frac{d\sigma_m}{d\varepsilon_m} \cdot V_m$$

where $(d\sigma_m)/(d\varepsilon_m)$ is the slope that the stress–strain curve of the *matrix* material alone would have at the same strain value as in the composite. At such strains, the matrix material is deforming plastically and so the slope of its stress–strain curve would be very low. Hence, to a first approximation, the modulus of the composite in stage (2) could be taken as $E_c = E_f V_f$.

Such a relationship indicates the necessity of using fibres of very high modulus for reinforcement.

The tensile strength of the composite is also a close function of the fibre strength. There is a minimum volume fraction of fibres V_0 above which the UTS of the composite will be reached at the same strain as the UTS of the fibres. This, however, does not imply that the UTS of the composite is the same as that of the fibres. Thus, the UTS of the composite is given as

$$\sigma_c = \sigma_f V_f + \sigma_{m'} V_m$$

where V_f is greater than V_0; σ_f is the UTS of the fibres alone; $\sigma_{m'}$ is the stress being carried by the matrix material when the strain in the composite is the same as the strain at the UTS of the fibres alone.

The relationship indicates that σ_c will increase with V_f and, in fact, σ_c/V_f plots are substantially linear. It is found that the stress being carried by the matrix is considerably lower than that being carried by the fibres and so this is true fibre reinforcement with the matrix transferring load on to the fibres.

It is obviously desirable to have V_f as high as possible to gain maximum strengthening but, in practice, fibre-to-fibre contact occurs if V_f approaches 0·8. If this does occur, σ_c tends to decrease, since there is then insufficient matrix material between the fibres to prevent the spread of cracks from fibre to fibre. The maximum useful fibre volume fraction is therefore about 80%. There is also a lower useful volume fraction of fibres. It is obvious that the addition of fibres must give a composite which has a higher UTS than that of the work-hardened matrix alone, otherwise there is no point in making the composite. The lowest volume of fibres V_L which will meet this condition is given by

$$V_L = \frac{\sigma_w - \sigma_{m'}}{\sigma_f - \sigma_{m'}}$$

where σ_w is the UTS of the work-hardened matrix material alone; V_L is usually higher than V_0 and varies with the properties of the fibre.

Discontinuous fibre reinforcement

The fibres here are still aligned in the direction of stressing but are not continuous throughout the length of the composite.

The UTS of such a composite is given by a relationship of the type

$$\sigma_c = \sigma_f V_f \left[1 - (1 - \beta) \frac{l_c}{l} \right] + \sigma_{m'} V_m$$

where l is the average fibre length; l_c is the fibre length in which plastic flow in the matrix could cause fracture of the fibre, i.e. the matrix deforms plastically and tends to shear over the fibre surface which is deforming only elastically—this drag effect could pull the fibre apart if there is enough fibre surface (length); β is a constant.

Comparing this relationship with that given for the UTS of continuous fibre-reinforced composites, i.e.

$$\sigma_c = \sigma_f V_f + \sigma_{m'} V_m$$

and realising that the term

$$\left[1 - (1 - \beta) \frac{l_c}{l} \right]$$

is always less than unity, it follows that discontinuous fibres give less strengthening effect than do continuous fibres.

Fibre-reinforced metals and their applications

Incorporation of the fibres into the metal matrix can be carried out by powder metallurgical methods of hot pressing and sintering, by liquid infiltration, by flame spraying of matrix on to the fibres or by electrodeposition. Processing temperatures must be low enough to prevent microstructural alteration of the fibres.

Titanium and its alloys have high strength with low density but have poor oxidation resistance and hot strength. Attempts have been made to improve the high-temperature properties by introducing 10–20% of cold-drawn molybdenum fibres. These are chopped to about 0·006 m lengths, incorporated by powder sintering followed by hot rolling of the sintered billet to develop the fibre. This material has improved creep properties as may be seen in Table 7.4.

Table 7.4

Material	UTS, MN/m²		Stress rupture life, h	
	473 K	923 K	733 K	813 K
Titanium	234	117	100	0·1
Ti+10% Mo	358	117	1000	100

Nickel, cobalt and iron-base alloys have also been successfully reinforced, using tungsten and molybdenum wires, giving much better high-temperature strengths.

Aluminium sheet has been strengthened, by incorporation of steel wires, by hot rolling continuous wires between sheets of aluminium. For example, by introducing about 10% of stainless steel wire, tensile strengths can be lifted from 62 to over 300 MN/m². Ceramic fibres such as glass and SiO_2 have also been used in aluminium alloys to give improved hot strength.

Fibre-reinforced components are small and necessarily expensive and so are only used for rather exotic purposes, e.g. for nuclear fuel plates, for reinforcing lead used in bearings, in rocket motors and for special turbine blading.

Fibre-reinforced plastics

Nearly any type of polymer can be used as matrix material. Thermosetting polymers with glass fibre reinforcement were the first to be used and are still widely employed to produce high-strength composites. The glass fibre is in the form of continuous filaments, woven mats or chopped strands. Thermoplastics may also be glass reinforced. Some use has also been made of whiskers of silicon nitride and aluminium oxide and of asbestos fibres.

If the matrix is a thermosetting polymer, the uncured resin is mixed with the fibres and then the material is cured to produce cross-linking as usual. Normal moulding techniques are used and for discontinuous fibre-reinforced thermosets, the material may be supplied as a pre-mixed moulding compound (dough-moulding) which contains uncured polymer, fillers and reinforcement. If continuous fibre reinforcement is needed, a continuous extrusion process may be used, the fibre and resin being extruded together through the moulding die. Glass-reinforced thermosets may contain anything up to 80% by weight of glass and are used for high strength–low density components such as boat hulls, car body

Table 7.5

Property	Thermosetting polymer	Polymer +50% glass
Density $\times 10^3$ kg/m^3	1·1–1·4	1·5–1·6
UTS, MN/m^2	35–70	100–170
E, MN/m^2	2070–4100	5500–14 000

panels, furniture, high pressure domes and shells. Table 7.5 indicates the improvement possible by glass fibre reinforcement.

Probably the greatest benefit of fibre reinforcement of thermosetting polymers is the improvement in toughness. For example, by introducing about 30% glass, the notched impact strength can be raised from 0·6 to 7·0 J.

Thermoplastic materials undergo no chemical change during moulding and so, in this case, the reinforcement technique is basically one of hot impregnating continuous fibres during the extrusion of rods of matrix material. The rods are then chopped into pellets and it is in this pelletised form that the material is usually supplied for the manufacture of mouldings. The glass content may be up to 70% by weight. With all thermoplastics, there is an increase in strength and modulus as a result of fibre addition but the effect on toughness depends largely on the original toughness. The impact strength of inherently brittle thermoplastics such as polystyrene is improved by reinforcement, but that of inherently tough material such as nylon may, in fact, be reduced. This behaviour is indicated in Table 7.6.

Table 7.6

Property	Nylon			Polystyrene	
	Unreinforced	20% glass	70% glass	Unreinforced	30% glass
Density $\times 10^3$ kg/m^3	1·14	1·31	1·45	1·05	1·28
UTS, MN/m^2	68·9	152	206·8	44·8	96·5
% El.	60	5	4	2	1·1
E, MN/m^2	2758	8274	21 374	2758	8274
Notch impact, J	1·36	2·7	2·03	0·40	3·4
Temperature to cause distortion under stress of 1820×10^3 N/m^2	340 K	>510 K	>510 K	357	380

17

Glass fibre reinforcement of plastics is obviously attractive but there are disadvantages. For example, plastic material begins to char and creep badly at temperatures around 500 K. Again, glass fibre, although it has high tensile strength, is not very stiff and so glass-reinforced plastic tends to bend easily under load.

Some of the more recent reinforced plastics make use of carbon or boron fibres in thermosetting resins. Such fibres are much stiffer than steel and so the composites have higher stiffness per unit weight than has steel. Such composites are being used for high-speed helicopter rotor blades and for compressor blades in small jet engines.

Composite materials produced from eutectic alloys

Artificially prepared fibre-reinforced composites present obvious difficulties in manufacture. Such difficulties arise in the production of the fibres in the first place, in the bonding of this constituent to the matrix and in the correct alignment of the fibres with respect to the applied service stress.

Such difficulties have stimulated research into the production of fibre-reinforced material direct from molten alloys by unidirectional freezing of eutectics. The eutectics most suitable for strengthening are those composed of intermetallic compounds. For example, the eutectic in the Al–Ni system is a mixture of Al and Al_3Ni. With normal or random solidification, the fully eutectic alloy in this system would give a UTS of about 96 MN/m^2 with 16% elongation. By unidirectional freezing to align Al_3Ni filaments in a matrix of Al, the UTS can be lifted to about 340 MN/m^2 with 2% elongation.

The amount of the strong intermetallic compound in any eutectic mixture is, of course, fixed for any given system and can be predicted simply by applying the lever law to the relevant equilibrium diagram. Therefore, unless intermetallic is artificially added, there is a limit to the degree of strengthening which can be produced.

Work has been carried out on the Cu–Cr and Al–Cu systems and there are indications that potential use might be made of refractory metal–carbide eutectic systems. For example, the tantalum–carbon system has a eutectic occurring at 8% carbon and containing about 30% by volume of extremely strong tantalum carbide Ta_2C. By persuading this Ta_2C to freeze in a rod-like form, the strength of the fully eutectic alloy can be lifted from 45 MN/m^2 to over 1000 MN/m^2.

Fig. 7.3 shows a photomicrograph of niobium carbide filaments

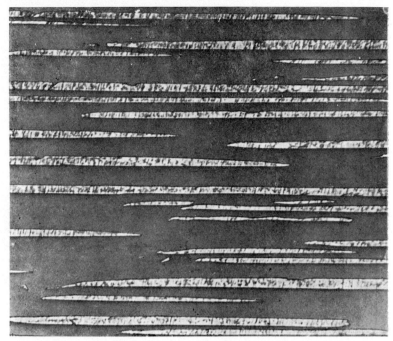

Fig. 7.3. Niobium carbide whiskers in a matrix of niobium (×2000) *(courtesy Drs. F. D. Lemkey and Michael J. Salkind, United Aircraft Corporation, USA)*

in a matrix of niobium, produced by controlled solidification of Nb–C alloy.

Apart from their high strengths, alloys containing directional eutectics have remarkable thermal stability. They can be heated to temperatures very close to the melting point without inducing major microstructural alterations. This stability of structure at elevated temperature produces stability of mechanical properties and so such materials may have potential as creep-resistant materials.

BIBLIOGRAPHY

BROUTMAN, L. J., and KROCK, R. H., *Modern Composite Materials*, Addison-Wesley (1967)
Composite Materials, Institution of Metallurgists (1966)
CRATCHLEY, D., 'Experimental aspects of fibre reinforced metals', *Metallurgical Review*, No. 95, Institute of Metals, London
Fibre Reinforcement of Metals, Ministry of Aviation, H.M.S.O., London (1965)

JONES, W. D., *Fundamental Principles of Powder Metallurgy*, Arnold (1960)
KELLY, A., and DAVIES, G. J., 'Principles of fibre reinforcement of metals', *Metallurgical Review*, No. 10, Institute of Metals, London (1965)
SCHWARZKOPF, P., *Powder Metallurgy*, Macmillan, New York
Scientific American, Vol. 217 (Sept. 1967)
TAYLOR, H. F. W., *The Chemistry of Cements*, Academic Press (1964)
TAYLOR, *Concrete Technology and Practice*, Angus and Robertson
THOMAS *et al.*, 'Properties of metal–ceramic mixtures', *Metallurgical Review*, Nos. 83 and 94, Institute of Metals, London (1963 and 1965)

Chapter 8

FAILURE OF ENGINEERING MATERIALS

The service behaviour of a material is not governed only by its inherent properties. The stress system on the material and the environment in which it is operating may be of overriding importance in fixing the life of a component, regardless of what its inherent properties are. This chapter, therefore, attempts to explain the response of materials to high and low temperatures, to suddenly applied stresses, to oscillating stresses and to corrosive atmospheres.

Most of the chapter is concerned with metallic materials, since the response of polymers and ceramics to stress and service environments has already been covered in the chapters devoted to those materials.

Failure of an engineering component may be defined in a number of ways. In many systems, changes in dimensions of the component parts cannot be tolerated and so failure would be deemed to have occurred if the material yielded. This is often the case with metallic components and nearly always the case with plastics. Against this, some systems, particularly those working at elevated temperature, are purposely designed to accommodate a certain amount of permanent deformation of the component parts. In many cases, however, failure is deemed to occur only when fracture occurs. This is always the case with ceramics and also refers to fatigue and brittle failures in metals.

It is therefore necessary to understand what fracture is and how it occurs. In a polycrystalline structure such as a metal or some ceramics, fracture occurs by the spread of a crack. The crack, because of the random arrangement of grains, usually starts at some point of stress concentration and then may progressively spread throughout the whole section. If plastic flow can occur, the tip of the crack is blunted and the crack may therefore be brought to a stop by this stress-relieving action.

The mode of fracture will therefore depend on how the crack

propagates and, in general, there are three important modes: (a) parallel slip planes shear completely over each other and voids open up as the planes then move apart; (b) parallel cleavage planes are pulled apart by a tension component; (c) grains move away from each other leaving intercrystalline voids.

The term 'ductile fracture' is used to describe the type of fracture which is preceded by a large amount of plastic flow, whereas 'brittle fracture' describes the appearance of a fracture which occurs without prior deformation. In both cases, the actual mode of fracture may be either shear, cleavage or intercrystalline.

To some extent, inherent material properties, such as cleavage strength, rate of work hardening, etc., will affect the way in which fracture occurs. For example, one would expect a brittle inter-crystalline type of fracture in a material which has low grain boundary strength but which work hardens rapidly.

However, there are a number of external factors which also affect fracture and failure and these are now briefly reviewed before a detailed account of actual failure mechanisms is given.

Grain size

This has a marked influence on both yield-stress and strain-hardening rate. In Chapter 2 it was shown that a plot of yield stress for BCC materials against $d^{-1/2}$ gave a straight line of equation

$$\sigma_y = \sigma_0 + ky\, d^{-1/2}$$

The fracture strength also varies with grain size in a similar way, but the slope ky is much lower.

With a fine-grained sample, deformation will be more uniform throughout the section and, in tension, plastic instability or necking would be less localised and would occur at higher strains. Such material undergoes more plastic deformation before fracture and the fracture is not likely to be intergranular.

Stress system and component geometry

Tensile components of stress are capable of opening out cracks and causing them to spread. Compressive stresses, on the other hand, will tend to close up any cracks already present. The resolved shear stress is about half the applied tensile or compressive stress if this is uniaxially applied and so shear fracture is not very likely under these conditions. Shear fracture may, however, occur in torsion since then the resolved shear and tension components are equal.

Because of the random arrangement and orientation of grains in a polycrystalline metal, it is impossible to apply, say, a true tensile stress throughout the whole sample. A uniaxially applied tensile stress will fall on different slip planes at different angles and so must generate a shear component.

Furthermore, the geometry of the sample may be such that an applied uniaxial stress actually becomes a multiaxial stress. A simple example of this occurs in the tensile test when plastic instability causes the sample to begin necking. As necking occurs, the section of the sample within the neck is trying to contract laterally but is prevented from doing this by restraint imposed by the rest of the sample. This restraint acts in a direction tending to be normal to the applied stress, i.e. it acts up the slope of the neck. A condition of triaxial stress is therefore set up.

Multiaxial stresses play a major part in fracture since, in their presence, plastic flow is inhibited and so a normally ductile material may fail in a brittle manner.

Consider Fig. 8.1: a uniaxial tensile stress σ_1 will produce a shear stress at 45° to itself and so slip could occur. Another tensile stress σ_2 applied at 90° to σ_1 would also resolve into a shear component at 45° to itself. There are now two planes of shear and these interfere with each other and restrict each other's movement.

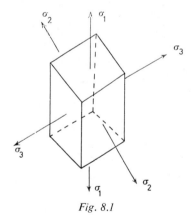

Fig. 8.1

A third tensile stress σ_3, at right angles to the other two, also generates its own shear stress and, in fact, if $\sigma_1 = \sigma_2 = \sigma_3$, the total resolved shear stress is zero and the slip planes cannot move. The material would then behave in a brittle manner since in such a case the yield strength would be the same as the fracture strength.

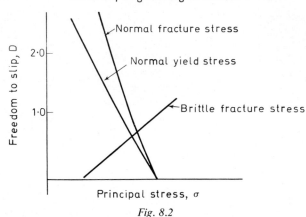

Fig. 8.2

The variations in behaviour of a material with variations in multiaxiality of stress can be illustrated, as shown in Fig. 8.2.

When $D = 0$, $\sigma_1 = \sigma_2 = \sigma_3$ and all are tensile. No slip can occur and the material fails in a brittle manner. When D is between 0 and 1, all stresses are tensile but σ_2 and σ_3 are less than σ_1. There is some slip possible but the fracture stress is still close to the yield stress. When $D = 1$, σ_1 is tensile and $\sigma_2 = \sigma_3 = 0$. This a condition of uniaxial tension.

When D is greater than 1, σ_1 is tensile while σ_2 and σ_3 are compressive. This type of combination obviously increases ductility. It is the reason, for example, why much larger deformations can be accommodated in, say, wire drawing than in a normal tensile test.

The geometry of the sample is important since any change in section can induce triaxial stress conditions, just as happens in the neck of a tensile specimen. The more pronounced the section the more severe is the restriction on slip. Notches or sharp section changes may therefore produce brittle fractures in an otherwise ductile material. This is, of course, typically the case in impact testing.

Strain rate

Dislocations moving through a crystal have to overcome obstacles, to accommodate themselves to other dislocations, and so on. It is therefore to be expected that such motion is time dependent. Increasing the strain rate may therefore cause restriction on plastic flow leading to brittle fracture since, as strain rate approaches dislocation velocity, the yield strength is brought closer to the fracture strength. In general, high strain rates set up multiaxial stresses because plastic flow is not given chance to relieve stress concentrations.

Temperature

Dislocation mobility is lower at lower temperatures. Decreasing the temperature will increase both yield and fracture stress but, as always, the yield stress increases most rapidly and so brittle failure may result. In BCC and CPH materials, the slip process is very sensitive to temperature and these materials usually become very brittle at some characteristic temperature. This ductile–brittle transition is discussed in more detail later.

Stress variation

A stressed component may begin to crack and fail at stresses which are well below its usual fracture strength and even below its normal yield strength, if the applied stress is oscillating. These are the so-called fatigue failures. In actual fact, it is only the *average* applied stress which is unexpectedly low. The real stresses which actually open out fatigue cracks are well above the normal fracture stress as a result of stress concentration. This topic of fatigue failure is discussed in more detail in the next section.

FATIGUE FAILURE

A component is liable to fracture without warning under repeated applications of stresses which it could support indefinitely if these stresses were static. This sort of unexpected failure is obviously of importance in such components as axles, shafts, boiler drums, aircraft cabins, rolls, forging tools, and so on.

In fatigue failure, a series of cracks, usually originating at some surface stress concentration, spreads gradually through the body of the material. This propagation stage can be of very long duration, depending on loading conditions, and involves very little change of overall dimensions. Fracture will occur suddenly when the remaining uncracked material is insufficient to support the applied load and this part of the fracture will be typically coarse. However, in spite of the brittle appearance of the final fracture, the material itself is not brittle. It is simply the overloading which produces this effect.

Since fatigue failure usually originates at some surface stress concentration, it is rarely a fault of the material. It is mainly a fault of design or of treatment. Typical surface stress concentrations are keyways, poor threads, lack of radius, grinding marks or cracks, quenching cracks, weld faults and so on. All these could be classified as 'metallurgical notches' and tend to introduce multiaxial stress conditions.

With a static stress, any stress concentrations would rapidly be relieved by plastic deformation occurring in a large volume of material around the stress concentrating point. With a cyclic stress, however, peak stress is present for a very short time during

Fig. 8.3. Fatigue failure (courtesy John Wiley and Sons Inc., New York)

each cycle and any plastic deformation which occurs is confined to a small volume of material near the stress concentrating point. One could imagine the concentrated stress being higher than the yield stress so giving rise to slip and work hardening. Eventually, all available ductility in the small volume of material at the root of the concentrating point is used up and a crack opens out. The tip of this crack itself is an efficient stress raiser and so further cycling will cause a repetition of the whole process with the eventual appearance of more cracks. As these cracks spread through the body of the component, the cracked surfaces find it possible to move slightly relative to each other under stress. This abrasive action leads to a rubbing or polishing of these surfaces and so a fatigue failure presents a characteristic appearance, as illustrated in Fig. 8.3. In this figure, the shell-like markings represent the inward progression of fatigue cracks and this surface is often quite smooth. One would look for the source of the failure at the focus of these markings. The final fracture is typically coarse and crystalline, representing sudden breaking due to overloading.

If, of course, the work hardening effect strengthens the material to such an extent that the concentrated stress cannot induce further slip, then fatigue failure will theoretically not occur. This will only be true if the loading conditions are kept the same.

DETERMINATION OF FATIGUE BEHAVIOUR

Fatigue behaviour is usually investigated by taking a number of carefully prepared specimens and subjecting them to endurance tests under a chosen stress system. The range of applied stress and the number of stress reversals to fracture are recorded. Further samples are tested at lower stresses and, finally, a stress range is reached at which fracture does not occur within say 10^7 to 10^8 reversals. Plotting the stress range S against the number of reversals N then gives the S–N curve. Frequently a log–log or semi-log plot is used. The first plotting point on these curves occurs at about 10^4 cycles.

The original test was the Wohler rotating beam test. This gives useful results but because component geometry and loading conditions play such a large part in determining fatigue behaviour, it is always wise to try and match the fatigue test to actual service conditions. Therefore, some typical forms of testing might be: (*a*) rotating beam with single or multi-point loading; (*b*) repeated flexure for sheet specimens; (*c*) push–pull tests; (*d*) oscillating torsion tests; (*e*) oscillating pressure tests.

The terms used in fatigue testing can be defined using Fig. 8.4.

$$\text{stress ratio} = \frac{\text{min. stress}}{\text{max. stress}}$$

$$\text{endurance ratio} = \frac{\text{fatigue strength}}{\text{UTS}}$$

Fatigue test results are often subject to considerable scatter and it is often desirable to analyse the results statistically.

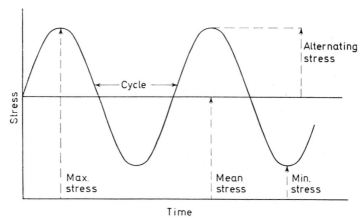

Fig. 8.4

The shape of the *S–N* curve varies according to the material being tested but metallic materials can be classified as: (*a*) materials which exhibit no ageing tendencies; (*b*) materials which strain age; (*c*) materials which age harden. These classifications are considered separately in more detail.

Non-ageing materials

Copper and its simple solid solution alloys are typical of this group. The *S–N* curves exhibit no limiting fatigue strength and so design must be based on limited life. Fig. 8.5 shows a typical *S–N* curve.

The fatigue strength in such materials is sensitive to variations in composition, grain size, degree of work hardening, testing temperature, and so on, as illustrated in Table 8.1.

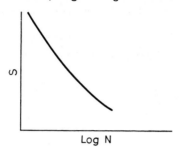

Fig. 8.5. Typical S–N *curve*

Table 8.1

Material	Condition	Grain size	Fatigue strength MN/m² for 10⁸ cycles	UTS MN/m²	Endurance ratio
Pure copper	36% cold reduction	4×10^{-5} m	123·5	336·6	0·37
Phosphor bronze	30% cold reduction	9×10^{-5} m	206·9	480·2	0·43
α brass	4% cold reduction	2.5×10^{-5} m	143·6	308·8	0·46

Strain-ageing materials

BCC materials in general and iron and steel in particular are typical here. A characteristic of the *S–N* curve is the appearance of a distinct fatigue or endurance limiting stress below which failure

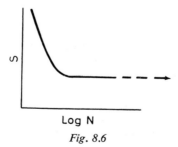

Fig. 8.6

will never occur, as indicated in Fig. 8.6. This introduces the possibility of designing for unlimited life. In strain-ageing materials, dislocations are pinned by atmospheres of interstitial atoms. A certain minimum stress is needed to produce freeing of dislocations and this is reflected in the occurrence of the fatigue limiting stress. Low carbon steels and some cast irons behave oddly in that the fatigue behaviour can be improved by treatments such as understressing or resting between periods of alternating stress. For example, a 30% increase in the fatigue strength of mild steel has been reported by stressing below the normal fatigue limit for $2 \cdot 5 \times 10^8$ cycles, then increasing the stress range gradually at intervals of about 15×10^6 cycles. This phenomenon is known as 'coaxing' and is again a reflection of the strain-ageing tendencies, i.e. time is given for the dislocations created by the initial slip process during fatigue to be pinned by Cottrell atmospheres and, hence, further slip is inhibited.

For smooth-surfaced samples of ferrous materials, the limiting fatigue strength is about 0·4–0·6% of the UTS, the upper limit being reached with sorbitic structures. This relationship breaks down if the material contains surface faults or if it is operating under corrosive conditions. Since this rather empirical relationship exists, any factor which raises the UTS should also raise the fatigue strength. Typical treatments would thus include alloying, heat treatments, cold working.

Non-metallic inclusions, such as oxides and nitrides, can drastically reduce fatigue strength if they are near the surface. This effect is particularly evident in high strength steels (in which ductility is inherently low) and major improvements in the fatigue strength of these steels have been produced by vacuum melting and casting.

Age-hardening materials

Aluminium alloys are typical here. Properly age-hardened alloys give high tensile strengths but, in spite of this, the fatigue strength is very little better than in an alloy which has been deliberately weakened by over-ageing. It seems that cyclic stressing of such alloys increases the rate of diffusion of solute atoms through the matrix and actually causes local over-ageing. The *S–N* curves for properly age-hardened alloys show no pronounced endurance limit. At very low temperatures where diffusion rates are negligible, the fatigue strength of an age-hardened alloy is much more in unison with its tensile strength, since the local over-ageing effect is prevented. This is illustrated in Table 8.2.

Table 8.2

Condition	Temperature K	Fatigue strength MN/m² at 10⁵ cycles	UTS MN/m²	Endurance ratio
Fully age hardened	290	277·9	579	0·48
	90	469·4	701	0·67
Over-aged	290	260·9	492·2	0·61
	90	398·3	560·5	0·70

EXPERIMENTAL OBSERVATIONS OF THE FATIGUE PROCESS

Some of the major observations can be listed as:

(1) Slip in fatigue occurs on the same slip planes as are used in static stressing and originates at stress concentrations.

(2) Most of the slip bands produced by fatigue stressing do not penetrate very deeply into the grains. A few bands, however, penetrate deeply and these eventually develop into cracks.

(3) The production of cracks from these deep slip bands occurs within 0·5 to 5% of the total fatigue life. The major part of the fatigue life is occupied in the propagation of these cracks and so periodic heat treatments, aimed at increasing fatigue life, are not very useful since such treatments cannot remove cracks.

(4) In many materials the slip bands produced during fatigue are associated with the extrusion of thin ribbons of metal, about 10^{-7} m thick and 10^{-5} m high, from the surface. These ribbons are extruded from the deep slip bands and hence fatigue cracks extend down these extrusions. A double-slip mechanism has been proposed for the formation of these extrusions. Consider a section of material having a dislocation source at A which is operated during the tension half-cycle to give a slip step on the surface at A^1 (Fig. 8.7(a)).

A second dislocation source B also operates during the same half-cycle, producing a second slip step at B^1 (Fig. 8.7(b)). This automatically displaces source A upwards to A_x.

The stress is now reversed into compression and the sources A_x and B now operate in the reverse direction. A_x operates and produces a second slip step on the surface at A^2 (Fig. 8.7(c)). This displaced slip produces an intrusion in the surface and

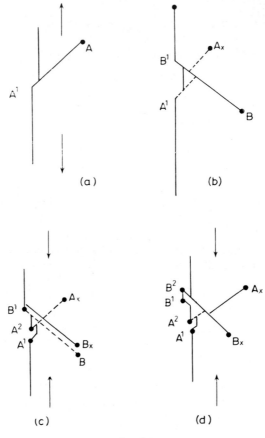

Fig. 8.7

also displaces source B to, say, B_x. Source B_x now operates under compression, but since this slip plane has been displaced an extrusion will be produced at B^2 (Fig. 8.7(d)).

It is feasible that these extruded ribbons on the surface are preferred sites for oxidation and that the movement of the ribbons into and out of the surface may transport oxide on to the slip planes, giving loss of cohesion. However, it has been observed that fatigue failure will still occur even in non-corrosive environments and so it seems more likely that the spread of a fatigue crack down the extrusions is simply a geometrical effect of the stress concentration at these points.

FACTORS INFLUENCING FATIGUE BEHAVIOUR

Presence of stress concentrations

The fatigue process originates at a stress concentration and it is probably fair to say that failure would never occur if all surface stress concentrations could be removed. The presence of stress concentrations is thus one of the major factors deciding fatigue behaviour.

A measure of the severity of stress concentration is given by the ratio

$$\frac{\text{max. local stress in region of a stress concentration}}{\text{average overall stress}} = K_t$$

In the case of a simple elliptical hole in a plate, for example, the stress $\sigma_{\text{max.}}$ around the hole is great erthan the average stress σ and is given by elasticity theory as

$$\sigma_{\text{max.}} = \sigma\left(1 + \frac{2a}{b}\right)$$

(see Fig. 8.8).

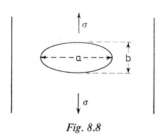

Fig. 8.8

Hence, in this case,

$$K_t = \frac{\sigma_{\text{max.}}}{\sigma} = 1 + \frac{2a}{b}$$

and so K_t is a factor by which the average stress must be multiplied in order to give the concentrated stress. A circular hole would give the least stress concentration and, obviously, the closer the shape approaches that of a slit or crack, the greater will be the concentrating effect; one would therefore expect that fatigue strength would be reduced as the severity of a stress concentration increases.

18

However, the reduction in fatigue strength which is actually found is less than the stress concentration factor and so a real fatigue strength reduction factor K_f is needed. This is simply the ratio of the fatigue strength in the absence of stress concentration to fatigue strength with a stress concentration.

The degree of agreement between K_t and K_f is given by the notch sensitivity factor q, where $q = (K_f - 1)/(K_t - 1)$.

Thus, as q increases from 0 to 1·0, the material becomes more sensitive to the presence of stress concentrations. q is not only a property of the material, since its value depends also on such factors as stress conditions, size of specimen, geometry of the stress concentration, and so on. For example, q increases with the size of the specimen and so stress raisers are more dangerous in large masses than in small. Again, an increase in tensile strength (which usually implies a reduction in available ductility) will increase the q value.

Mean stress

The fatigue strength of a material is quoted as that value of alternating stress which does not cause fracture after a given number of stress reversals. This, however, is not a complete statement since there may be a mean or standing stress superimposed on the alternating stress. The possible conditions are indicated in Fig. 8.9.

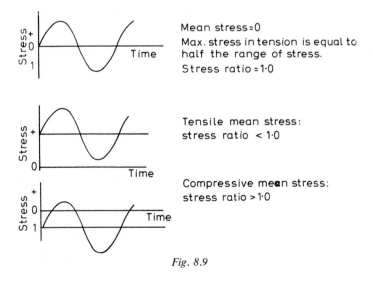

Mean stress = 0
Max. stress in tension is equal to half the range of stress.
Stress ratio = 1·0

Tensile mean stress:
stress ratio < 1·0

Compressive mean stress:
stress ratio > 1·0

Fig. 8.9

Various empirical relationships exist between fatigue strength and imposed tensile mean stress. Tensile mean stresses are of major importance since these open out any fatigue cracks, and their influence can be represented on a so-called *R–M* diagram. This is simply a plot of *R*, the range of alternating stress (2×alternating stress), against *M*, the mean tensile stress. A number of relationships have been proposed, e.g. by Gerber, by Goodman and by Soderberg, and these are illustrated in Fig. 8.10.

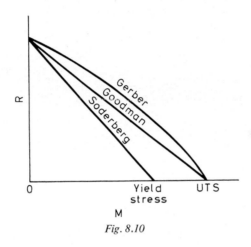

Fig. 8.10

The various relationships may be expressed as

$$\text{Goodman:} \quad \sigma_a = \sigma\left(1 - \frac{\sigma_m}{\sigma_u}\right).$$

$$\text{Gerber:} \quad \sigma_a = \sigma\left(1 - \left[\frac{\sigma_m}{\sigma_u}\right]^2\right).$$

$$\text{Soderberg:} \quad \sigma_a = \sigma\left(1 - \frac{\sigma_m}{\sigma_y}\right).$$

where σ_a is the alternating stress $(R/2)$ associated with a mean tensile stress σ_m; σ is the alternating stress at zero mean stress; σ_u and σ_y are the UTS and yield stress, respectively.

The general impression from all these relationships is that a tensile mean stress will drastically lower the fatigue strength. However, this statement must be qualified because its validity depends on the surface finish of the test samples, and the type of

18*

alternating stress. This is illustrated by considering Figs 8.11 and 8.12.

Fig. 8.11 refers to smooth-surfaced samples tested under shearing fatigue stress. It is a plot of the ratio

$$\frac{\sigma_f}{\sigma_0} = \frac{\text{fatigue strength with mean stress applied}}{\text{fatigue strength without a mean stress}}$$

against the ratio

$$\frac{\sigma_{\text{max.}}}{\sigma_y} = \frac{\text{max. stress of fatigue cycle}}{\text{yield strength}}$$

The results are subject to considerable scatter but it would appear that the presence of a tensile mean stress has no influence on fatigue strength, at least until the applied fatigue stress equals the yield stress.

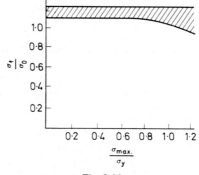

Fig. 8.11

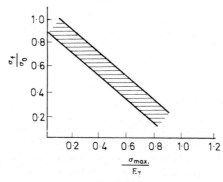

Fig. 8.12

Fig. 8.12, however, refers to samples containing surface imperfections. It is a plot of the ratio σ_f/σ_0 against $\sigma_{max.}/E_T$, where E_T is the modulus of rupture in torsion.

In this case, the presence of a tensile mean stress definitely lowers the fatigue strength. The last two figures refer to tensile mean stresses superimposed on alternating shearing stresses. Various other combinations have been investigated and the results can be summarised thus:

(1) A tensile mean stress superimposed on an alternating tensile stress will reduce the safe range of alternating stress no matter what the surface condition of the sample.

(2) A compressive mean stress is usually found to be beneficial since, for samples having surface imperfections, it will increase the safe range of alternating stress. This obviously suggests one way of reducing the risk of fatigue failure.

Influence of microstructure

At normal temperatures, the grain boundaries in crystalline materials are efficient barriers to dislocation motion and so fine-grained material will tend to inhibit slip and the formation of microcracks. Treatments designed to refine grain size may therefore improve fatigue resistance.

In the case of multi-phase materials, grain size is not as important as the type, shape and distribution of the precipitated phases. The important factor is the mean free path from one particle to the next, since it is within this free path through the softer matrix that slip occurs. A sorbitic steel, for example, would have better fatigue resistance than the same steel in which the carbide was in the pearlite condition. Shape is also of importance, particularly if the precipitated particles are near the surface. One could cite here the poor fatigue properties of flake graphite iron as compared with nodular iron.

Influence of temperature

Slip is less easy at lower temperatures. This arises because the total stress acting on a dislocation is partly due to the applied mechanical stress and partly to the thermal stress arising from atomic vibrations. It is generally found, therefore, that fatigue strength is better at lower temperatures, as indicated in Table 8.3.

Table 8.3

Material	Endurance ratio		
	293 K	230 K	83 K
Carbon steel	0·43	0·47	0·67
Alloy steel	0·48	0·51	0·58
Aluminium alloy	0·42	0·50	0·59

At temperatures well above room temperature, fatigue behaviour is secondary to creep behaviour and, in fact, the creep rupture strength is generally less than the fatigue strength. However, with most materials, as one would expect, fatigue strength is lowered as temperature increases. Strain-ageing materials such as carbon steel may behave peculiarly in that they will exhibit an improvement in fatigue strength between about 373 K and 620 K. This is the temperature range in which any dislocations created by straining would rapidly be made inoperative by diffusion of interstitial atoms into them.

Influence of frequency of stress cycle

At normal temperatures, the fatigue life of most materials is not much influenced by the speed of cycling, at least in the range 500–10 000 c/min. What little effect there is, is usually a slight decrease in fatigue strength as frequency decreases but this may be due to the fact that, at lower frequencies, more time is spent before failure and, hence, there is more opportunity for surface oxidation and corrosion. At elevated temperatures, this effect becomes more marked and the fatigue life tends to depend on the total time of testing rather than on the number of cycles.

Influence of residual stress

Residual stress in a component arises either mechanically, e.g. as a result of plastic deformation, or thermally, e.g. as a result of inhibited thermal contraction. Local surface irregularities will usually be associated with residual stress since plastic deformation may occur around these stress concentrations during manufacturing processes, even though the bulk of the material has not been plastically deformed. In general, a residual tensile stress will lower fatigue strength while residual compressive stress is usually beneficial.

Fusion welds contain notoriously high residual tensile stresses, resulting from inhibited thermal contraction, and so one would expect an improvement in fatigue resistance of welded components after post heat treatment. In fact very little improvement results. The reason for this is the overriding influence of the weld profile and it is the stress concentrating effect of this which swamps all other factors. Table 8.4 illustrates this effect.

Table 8.4

Material	Angle θ	Fatigue strength for 2×10^6 cycles MN/m²
Mild steel	0°	247·1
Mild steel	30°	177·6
Mild steel	50°	92·6

Influence of environment

Corrosion and oxidation produce surface roughening and this can lead to a drastic reduction in fatigue strength. This corrosion fatigue is particularly troublesome in marine equipment, boiler and superheater tubes, turbine and pump components. The process has certain features which distinguish it from normal fatigue:

(1) With materials which normally exhibit a well-defined endurance limit, the presence of a corrosive environment removes this limit and so designing for unlimited life is no longer possible.
(2) Corrosion fatigue strength is markedly dependent on the frequency of the alternating stress.
(3) Raising the UTS will normally raise fatigue strength but this does not always pertain under corrosive conditions.

Because of the drastic effect of corrosion on fatigue behaviour, it is usually found that the corrosion fatigue life of a component is primarily dependent on its corrosion resistance and not on its fatigue strength.

The accelerating effect of corrosion on fatigue failure is the result of a number of factors. For example, work-hardened material always undergoes preferential corrosive attack and such material is present at the root of a fatigue crack. Again, the presence of alternating stress and strain produces constant rupture of any protective oxide films on the material and so bare material is always exposed to corrosive attack.

A peculiar case of corrosion fatigue occurs as a result of *fretting corrosion*. This sort of corrosion is often found in components which are clamped together tightly. The components are not supposed to move relative to each other but, in fact, do. The relative motion to give fretting corrosion need only be about 2×10^{-8} m. Such contacting metal surfaces only make true metal–metal contact at a few high spots and this true area of contact is much less than the apparent area. As a result, the stress on these contacting spots is very high and they may weld together. During slight relative motion, these welded junctions are sheared and what we usually call wear results from the ploughing of these high spots through the matrix. The wear particles produced are very small, and oxidise rapidly. The oxides cannot escape from between the tightly fitting surfaces and so penetrate into the metal giving surface roughening. It is this roughening which can lead to the production of fatigue cracking over the contacting surfaces.

Most of the influencing factors already discussed can be investigated using normal fatigue tests. Such tests give S–N curves on which the first plotting point will occur at about 10^4 cycles or more. The conventional S–N curve cannot, therefore, tell us anything about the fatigue stress which would be given at endurances less than about 10^4 cycles. However, in some components such as pressure vessels and aircraft landing gear, the total service life may involve only a few thousand stress cycles and it may be that the material can accommodate stresses which are well above the fatigue limit as established by long time tests. It would be dangerous simply to extrapolate the conventional S–N curve back to, say, 100 cycles and so an adjusted form of testing is needed, using a small number of stress reversals at high amplitudes of stress or strain.

This immediately raises a problem, since at these high testing stresses there is often considerable plastic deformation during each cycle. This means that the deformation is non-Hookean and stress is no longer proportional to strain. It is therefore necessary to distinguish, under these conditions of testing, between the resistance to alternating stress and the resistance to alternating strain. These will be considered separately.

Influence of small numbers of stress reversals at high stress amplitude

The information made available by tests at high stress amplitude is conveniently summarised, as indicated in Fig. 8.13. This plot refers to axial fatigue stressing and has been drawn so as to include the results from a wide variety of metallic materials. In spite of this, the spread of the results is remarkably small and the curve indicates that the fatigue strength at low endurances is closely

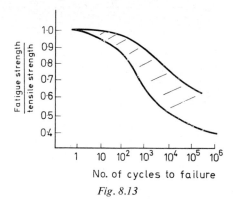

Fig. 8.13

related to the normal tensile strength. For example, an endurance ratio of unity should occur at a quarter-cycle and this would correspond to a normal tensile test. As the number of cycles increases, the fatigue strength becomes lower than the tensile strength but this occurs in a regular manner.

Influence of small numbers of stress reversals at high strain amplitudes

Fig. 8.14 is a plot of strain against endurance and again includes results from a variety of metallic materials. The scatter band is again remarkably narrow and it is clear that the fatigue strength in this case is not related to the tensile strength but to the ductility. This relationship pertains up to about 2×10^4 cycles. Above this, the normal relationship between fatigue strength and tensile strength reappears. It would appear, therefore, that since below

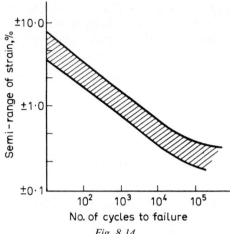

Fig. 8.14

about 10^4 cycles endurance is mainly a function of strain amplitude (ductility properties), nothing is to be gained by the use of high-strength materials under such conditions. For example, a high-strength steel and a light alloy, both having the same ductility, would both have similar fatigue lives under conditions involving a few cycles of high strain amplitude.

PREVENTION OF FATIGUE FAILURE

Much can be done in the selection of material, design and pre-treatment stages to reduce the risk of fatigue failure.

In *material selection*, one is usually guided by the fact that there is a fairly clear relationship between fatigue strength and tensile strength. This concept is, however, subject to limitations since

(*a*) the use of a high strength material will give very little improvement in fatigue strength in conditions of fretting corrosion, corrosion fatigue, or where the material is subject to a small number of stress reversals at high strain amplitudes,

(*b*) the relationship between fatigue and tensile strength fails with very high-strength materials since these tend to be notch sensitive. In fact, the normal endurance ratio of 0·5 pertains with steels, for example, only up to about a UTS of 1250 MN/m². Better results can be obtained from such steels by vacuum melting and casting since this removes oxides and nitrides which would otherwise appear as non-metallic inclusions and might act as stress raisers. Table 8.5 indicates the effect of vacuum degassing on the fatigue properties of a carbon–chromium ball bearing steel.

Table 8.5

Material	*Production process*	*Fatigue strength* MN/m^2
Carbon–chromium steel	basic electric	587
	vacuum melted	850

Ausformed steels give high tensile strengths with better than usual ductility, and fatigue strengths as high as 1160 MN/m^2 for 10^7 cycles have been reported. Such steels are not, however, economical for normal applications. A similar sort of situation arises with aluminium alloys since, again, there is only a slight increase in fatigue strength beyond a tensile strength of about 390 MN/m^2. Below this, the endurance ratio is about 0·5. High strengths in such alloys are usually the result of age hardening, and alternating stresses cause local over-ageing with considerable loss of strength.

The notch sensitivity of the material must usually be taken into account and tests could be made to assess the notched fatigue strength. Notch sensitivity is, however, dependent on external factors such as component size as well as on material properties and it is usually wise to assume in design that the material is fully notch sensitive.

The proposed surface finish of a component could influence selection procedure since it is often the case that a material of low strength but with a surface treatment will behave better in fatigue than will a high-strength untreated material.

Vibration may be an accelerating factor in fatigue failure and the damping capacity of the material may be of importance. It may therefore be better to use a weaker material (more ductile) of high damping capacity than a more expensive high-strength material which may have a higher fatigue strength but which may also fail sooner because of the imposed vibration.

The design of a structure or component operating under alternating stress conditions should aim to eliminate, as far as possible, any surface stress concentrations. This usually calls for determination of the stress distribution over the surface by calculation for simple shapes or by experimental stress analysis. The design might also specify some form of surface treatment. Such treatments have a marked influence on fatigue behaviour and may be more important than the inherent fatigue strength of the material.

Surface treatments can influence fatigue behaviour in the following ways: (*a*) by increasing the intrinsic hardness and tensile strength of the outer layers, hence increasing fatigue strength—e.g., cold peening, cold rolling; (*b*) by introducing compressive surface stresses or removing surface tensile stresses—residual compressive stresses are induced by surface hardening, plating, tensile overstrain, etc.; (*c*) by improving surface finish—plating and surface-hardening treatments are particularly beneficial since they induce surface compressive stresses as well as giving smooth surfaces; they may also improve corrosion resistance.

BRITTLE FRACTURE—DUCTILE-BRITTLE TRANSITIONS

Some metallic materials, which are normally considered to be quite ductile, can fracture catastrophically in a completely brittle manner. This disturbing possibility is a feature of BCC and CPH materials but does not occur with FCC materials. The tendency towards brittle fracture is increased at low temperatures and at high strain rates. Such high strain rates are usually introduced by the presence of notches or stress concentrations and, in fact, the fractures obtained during impact testing and in the final stages of fatigue are typical brittle fractures.

Brittle fracture is characterised by very high speed propagation of a crack (up to 1200 m/s in steel) at low overall stresses, the fracture occurring with very little prior plastic deformation. The fractured surfaces are bright, coarsely crystalline and often exhibit chevron or 'V'-shaped markings. The material is not, however, fundamentally brittle, since tensile tests on the fractured pieces indicate quite normal elongation values.

Attention was focused on this problem during the last war when a number of merchant ships developed major hull cracks after only a short time in service. A few of these ships broke up completely immediately after launching. A view of this catastrophe is given in Fig. 8.15.

These ships had been fabricated by welding instead of by the usual riveting process. This does not, however, mean to say that welding was responsible for the failures. Brittle fractures do occur in riveted structures but the crack spreads only to the edge of a single unit or plate. Welding is significant in that it produces

Fig. 8.15

continuous structures and so a crack, once started, has the opportunity to travel large distances.

Many of the ship failures occurred when the air temperature was lower than usual and were found to originate at stress concentrations such as hatch corners, or weld faults. The cracks, however, rarely followed the line of the weld, and so although poor welding technique might be responsible for initiating a crack, the presence of weld metal certainly does not aid in propagating the crack.

Considerable research has been carried out on this problem since that time because, apart from affecting our main constructional material, mild steel, it also affects CPH metals and the newer structural metals such as niobium, titanium, chromium and molybdenum.

There appears to be no simple remedy for brittle failure, but enough is known to be able to understand the mechanisms of failure and to guard against it.

Since in brittle failure, a normally ductile material is behaving in a completely brittle manner, it may be profitable to consider the Griffith criterion for crack propagation in brittle materials (see page 235). Griffith carried out his work on glass, a non-crystalline material in which the slip process is impossible. He postulated the existence of microcracks and his theory deals with the spread of elastic cracks through the material. At the root of such a crack in glass, slip movement is not possible and so the atomic bonds will be stretched in varying degrees of tension. As the crack spreads, bonds must be broken and stress is thrown on to bonds further forward. The work done in stretching and breaking these bonds can then be equated with the surface energy of the fractured faces. For a crack to spread in this way, the theoretical strength of the material needs to be exceeded only at one point at a time and so the actual applied stress needed to cause fracture will be much less than the theoretical stress, i.e. the crack is acting as a stress concentration. The applied stress produces elastic strain energy in the material:

$$E \times \frac{(\text{strain})^2}{2}$$

Thus, as the crack spreads, elastic strain energy is used up and surface energy is created. Griffith suggested that a state of balance could exist between the energy needed to produce two new surfaces by fracture and the strain energy in the material. This energy balance is usually expressed as

$$\sigma = \left(\frac{E\gamma}{c}\right)^{1/2}$$

where σ is the smallest tensile stress needed to spread a crack of length C; γ is the surface energy of the fractured faces; E is the elastic modulus

The implication of this is that the crack will spread if the reduction in strain energy produced by increasing C is greater than the increase in surface energy produced by the increase in surface area of the cracked faces. It also implies that there is a critical crack size for propagation to occur.

The Griffith theory has been verified for completely brittle materials such as ceramics but it cannot be applied directly to metals. There are two main reasons for this.

(1) Metals which fail in a brittle manner do so only below a certain critical temperature. None of the terms in the Griffith equation is particularly sensitive to temperature.
(2) When metals fracture in a brittle manner, the fracture is always preceded by a small amount of plastic deformation and so the cracks are not purely elastic.

There must therefore be other factors operating in metals besides the Griffith mechanism, and it is possible to modify the Griffith energy balance by introducing a plastic work factor to account for the additional energy needed to produce local plastic deformation along the surfaces of the cracked faces.

The implication of this is that slip dislocations must play some part even in a brittle fracture process, and it is now believed that these dislocations actually originate the stress concentrations which can then open out into microcracks. A Griffith-type energy balance may therefore exist between the stress needed to cause dislocation motion (the yield stress) and the stress needed to cause growth of microcracks (surface energy). If the yield stress is at a higher level than the surface energy then the material will fracture before it yields and so will be brittle.

A simple model for the formation of microcracks from slip dislocations is illustrated in Fig. 8.16.

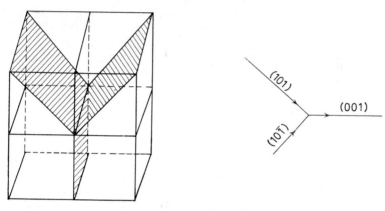

Fig. 8.16

In this model, dislocations in a BCC lattice are moving down the {110} slip planes. With some conditions of temperature or rate of loading a pile-up may occur at the intersection and one can then imagine the pile-up stress being transferred on to the (001) cleavage plane. The cleavage planes, because of this wedging action, come under a tensile stress and so a microcrack could open out. A dislocation pile-up of this nature would obviously be favoured if large numbers of dislocations were suddenly produced. This is exactly what happens in BCC materials when they yield (see p. 59). This avalanche of dislocations produced at the onset of yielding could feasibly lead to pile-up and microcrack formation at high strain rates. Such a crack will spread if the strain rate is so high that there is insufficient time for stress relaxation to occur by blunting of the crack tip.

The key to this problem seems to be in understanding those factors which could cause dislocation motion to be less easy than usual. Any such process will, of course, raise the yielding stress since by definition this is the stress at which dislocation motion first occurs. If the condition is reached at which the yield stress is higher than the stress needed to create new internal surfaces then a crack will propagate rapidly and brittle fracture will result.

In BCC materials, at least, there is obviously some relationship between those factors responsible for the yield drop observed in these materials and the brittle fracture behaviour. In these materials, the yield stress and grain size are related via the Petch equation (see p. 59)

$$\sigma_y = \sigma_0 + kyd^{-1/2}$$

σ_0 is usually quoted as a friction stress opposing the movement of free dislocations through the lattice. The parameter ky is probably related to the stress needed to free or create the dislocations in the first place. A general equation can be derived which expresses the conditions which will produce a microcrack at the yield stress instead of producing plastic flow (yielding). The equation can be stated in the form

$$q\mu\gamma = ky(\sigma_0 d^{1/2} + ky)$$

q is a stress concentration factor such that the axial tensile stress in a notch at yield is raised to $1/q \times \sigma_v$; q thus equals $1 \cdot 0$ in the absence of a notch and it has a value of $\frac{1}{3}$ in the presence of a stress concentration; μ is the shear modulus; γ is the surface energy of a crack.

If conditions are such that the left-hand side of the equation is greater than the right-hand side, then yielding would be expected to occur before fracture and the material would be ductile. The right-hand side of the equation is of major importance since it contains those terms which are most variable. Hence, brittleness would be favoured by an increase in any of the terms in this side of the equation.

σ_0 is raised by decreasing the temperature, by strain hardening or strain ageing, by the presence of fine precipitates, by increase in strain rate, by irradiation or alloying and by the presence of impurities in interstitial solid solution.

ky is independent of temperature and strain rate and probably independent of composition also, although there is some evidence that it is reduced in ferrite by manganese additions.

$d^{1/2}$ is increased as grain size increases.

We have, therefore, a list of those factors which will induce the transition from ductile to brittle behaviour and some of the more important of these factors can now be considered in more detail.

INFLUENCE OF STRAIN RATE AND TEMPERATURE

The force on a dislocation is a combination of the applied mechanical stress and the stress due to thermal vibration of the ions. It is therefore to be expected that slip will be more difficult at low temperatures and that the yield stress will increase as temperature drops.

A dislocation moving through a metal lattice experiences a resistance to its motion, since the metal ions in the dislocation must move through the fields of force surrounding the other ions in the lattice. This resistance or lattice friction is termed the Peierls–Nabarro force. The σ_0 parameter in the Petch relationship is partly the result of this inherent lattice friction and partly the result of lattice strains introduced by solid solution formation, etc. It is the Peierls–Nabarro component of σ_0 which is temperature sensitive, and in BCC and to a lesser extent in CPH materials, this friction stress increases very rapidly as temperature drops. In FCC materials, because of the different arrangement of ions in the lattice, the Peierls–Nabarro force is not so sensitive to temperature. There is the possibility, therefore, in BCC and CPH that at some temperature the yield stress coincides with the fracture stress and so brittle failure occurs. Because of the very rapid increase in the Peierls–Nabarro force as temperature drops, we should also expect the transition from ductile to brittle behaviour to be rather sharply defined.

BCC and CPH materials are also peculiar in another respect, since both impose inherent limitations on dislocation movement. In CPH systems, for example, there is only one major set of slip planes, the {0001} basal planes, and these are also the cleavage planes. The stoppage of slip by pile-up dislocations can therefore lead directly into fracture.

In BCC systems, we have already seen the possibility of cleavage resulting from dislocation pile-up on intersecting slip planes and, as usual, such pile-up would be favoured by low temperatures or high strain rate. Another factor here is that the inherent dislocations are strongly pinned by Cottrell atmospheres of interstitial impurities. Yielding is then equated with the creation of an avalanche of fresh dislocations and it has been shown that BCC materials can support stresses much higher than the normal yield stress for short

19

periods of time without undergoing plastic deformation. Mild steel, for example, will withstand $2\sigma_y$ without yielding for about 0·00001 s if the stress is suddenly applied. This 'delay' time is, of course, the time needed for the creation of fresh, mobile dislocations but it is at once obvious that the yield stress is very sensitive to strain rate and that impact loading, for example, could cause brittle failure. This delay time is temperature sensitive and increases as temperatures decreases.

The impression is therefore gained that in BCC and CPH systems, the production or movement of dislocations is attended with some difficulty and that these processes are particularly sensitive to temperature and strain rate. One could imagine that, above some critical temperature, the time needed for dislocation generation and motion would be negligible and so the material would be ductile at any strain rate. Below such a temperature, the time needed would be appreciable and so, depending on the strain rate, the material could appear to be brittle. This introduces the concept of a ductile–brittle transition temperature.

As one would expect, the actual transition temperature is very dependent on method of testing, geometry of test piece, the imposed stress system, and so on. Mild steel, for example, would show a transition from ductile to brittle behaviour at 100 K using a normal tensile test but this may be raised to nearer 273 K by using a notched test piece.

Normally, some form of notched bar impact test is used since the conditions for producing brittle failure are most accentuated in the presence of a notch at high strain rates. The results of such tests are not, of course, quantitatively applicable to service structures since the geometry of the systems will be different.

The variable in the test is temperature and the results are plotted in the form shown in Fig. 8.17.

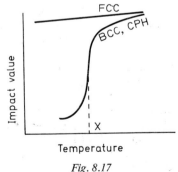

Fig. 8.17

The ductile–brittle transition temperature *Tc* may be taken as that temperature at which the fractured faces exhibit 50% bright, crystalline brittle fracture and 50% ductile fracture, i.e. point *X*.

Curves of this nature indicate that the transition in BCC and CPH materials is quite sharp. Obviously, the material becomes dangerous to use if *Tc* coincides with operating temperature. *Tc* varies widely with the stress system imposed and so care is needed in design and fabrication to avoid stress concentrations since these, in effect, locally raise the strain rate.

Table 8.6

	Material	*Tc* K
FCC	Copper	no embrittlement
	Nickel	no embrittlement
	Aluminium	no embrittlement
BCC	Iron	220–273
	Chromium	about 470
	Tungsten	450–620
	Molybdenum	about 300
CPH	Magnesium	230–273
	Zinc	230–273
	Titanium	about 273

Table 8.6 gives some values for *Tc* but these are necessarily approximate since the actual *Tc* depends on factors such as grain size, purity, metallurgical condition, geometry, surface finish, and so on.

INFLUENCE OF GRAIN SIZE AND COMPOSITION

In a polycrystalline sample, dislocations will run down a slip plane which is considered as extending right through a grain from boundary to boundary. If a pile-up occurs on such a slip plane, the number of dislocations in the pile-up is obviously a function of the length of the slip plane, i.e. of the grain size. Hence, the elastic stress concentration at the head of a pile-up will increase as grain size increases and so has more chance of opening out a micro-crack. It is then to be expected that refining the grain size would lower the value of *Tc* and so reduce the possibility of brittle fracture. A linear relationship actually exists between *Tc* and $\log_e d^{-1/2}$, as shown in Fig. 8.18.

19*

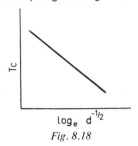

$$\text{log}_e \ d^{-1/2}$$

Fig. 8.18

Decreasing grain size will give an increase in yield strength and, from what has been said before, one might expect this to increase the tendency towards brittle failure since the yield stress is brought closer to the fracture stress. However, the important factor here is the reduction in the length of slip pathways produced by refining grain size, i.e. the reduction of the term $d^{1/2}$ in the general transition equation on page 282.

Treatments designed to refine grain size will therefore lower Tc and, in the case of steel, these might include thermomechanical treatments, normalising, addition of grain-refining elements such as aluminium or niobium, etc.

Chemical composition is also important in fixing Tc since it can influence both σ_0 and $d^{1/2}$ in the general transition equation. Alloying elements which enter into solid solution will raise σ_0 and hence raise the yield stress closer to the fracture stress. If they do this without producing a corresponding decrease in grain size, then in general they will raise Tc and so increase the tendency to brittle fracture. Alloying elements which produce coarse, brittle precipitates will also raise Tc. As regards mild steel, foreign elements which enter into solid solution in ferrite include carbon, silicon, manganese and nitrogen, while those giving brittle precipitates are carbon (Fe_3C), sulphur (FeS) and phosphorus (Fe_3P). The influence of carbon content on Tc is illustrated in Fig. 8.19. The raising and broadening of the Tc range is due here to the increase in the amount of cementite in the microstructure.

Fig. 8.20 indicates the general influence of common alloying elements in steel. The beneficial grain-refining effects of manganese and nickel are evident and it is for this reason that a high Mn : C ratio is aimed at in the production of notch-tough mild steel.

Elements which enter into interstitial solid solution usually have the most severe effect on Tc since these are the ones responsible for the formation of Cottrell atmospheres. Nitrogen, hydrogen, oxygen and carbon in interstitial solution in metals such as iron, chromium, molybdenum will all raise Tc.

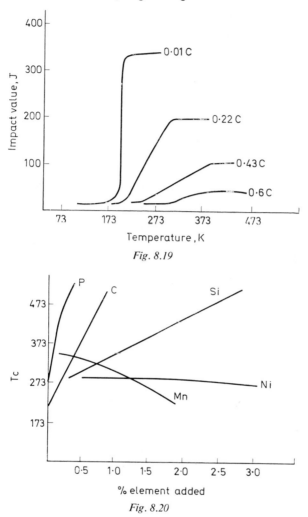

Fig. 8.19

Fig. 8.20

WELDING AND BRITTLE FAILURE

There is no reason to suppose that the use of welding as a means of construction will automatically introduce an increased tendency to brittle failure. It is, however, obvious that a brittle crack starting in a welded structure will be able to travel further and cause more damage than would a crack in a riveted structure.

It is necessary, when considering welding, to separate crack initiation and crack propagation. The propagation or spread of a crack in a welded construction rarely occurs via the weld metal itself. In fact, weld metal usually has a lower Tc value than the parent metal. One assumes here that the weld metal has been deposited with adequate protection from pick-up of atmospheric gases.

Crack initiation, however, may be a direct result of welding, inasmuch as weld faults such as slag inclusions, porosity, lack of penetration, undercut and hot cracking all constitute stress concentrations and so will tend to raise Tc. Such faults are, of course, under the control of the operator and can be avoided.

Fusion welds usually contain quite high residual tensile stresses and although one would expect these to increase brittle failure tendencies, the statistical evidence seems to indicate that such stresses have little influence. In spite of this, it would seem to be good policy to remove such stresses.

CREEP IN METALLIC MATERIALS

Creep is one aspect of viscoelastic behaviour and can be defined as the time-dependent deformation which accompanies the application of stress to a material, particularly at elevated temperatures.

Creep in polymers, as was indicated in Chapter 5, is very temperature sensitive and the viscous component of deformation is very evident in such materials. As a result, they exhibit high rates of deformation at relatively low temperatures and stresses.

Creep in metals, because of their total crystallinity and different bonding mechanism, is somewhat different from this. In metals, creep deformation is usually essentially plastic but it may become viscous in nature at higher temperature or stress levels. Even then, however, the deformation does not involve pure viscous flow, since the rate of deformation is not a linear function of stress at any given temperature.

Creep in metals is important with any component which is to operate at elevated temperatures, e.g. furnace plant, boiler plant, turbine blades, high-temperature engine components. The stresses which cause the dimensional changes are below the static yield stress and operating temperatures are, in any case, below the normal recrystallisation temperature. Hence, one would not expect dimensional changes to occur. However, the key factor here is the time over which the stress is applied, long-time loading causing permanent deformation at much lower stresses than would be needed in a short-time test.

General characteristics of creep

The standard creep test measures the strain or elongation of a loaded sample as a function of time at a constant temperature. If a constant load is used, the stress will not, of course, remain constant, since the sample will undergo elongation. However, constant loading is more representative of service conditions.

Some typical creep curves are illustrated in Fig. 8.21. In such creep curves for real materials, there may be four distinct regions although not all materials exhibit all these stages. A material which does exhibit all possible aspects of creep behaviour is graphite.

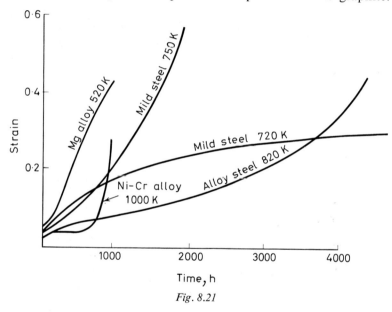

Fig. 8.21

This is a crystalline material and can be used as a model for metallic creep. A creep curve for graphite is illustrated in Fig. 8.22. The four stages of the deformation are:

I   — instantaneous extension produced as soon as the test load is applied. The material work hardens up to the applied stress and so this deformation is partly elastic and partly plastic.

P   — primary or transient creep stage during which further work hardening occurs. The flow stress tends to become higher

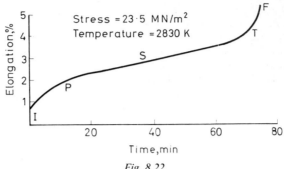

Fig. 8.22

than the applied stress but this work-hardening effect does not stop the creep, since at the temperature being used, diffusion and thermal activation can occur to produce softening, i.e. the slip process going on during stage *P* generates and jams dislocations but the operating temperature allows some freeing of dislocations by recovery. The recovery effect is, however, less than the work-hardening effect and so the creep rate during this stage decreases logarithmically and eventually leads into

S — a steady state or secondary creep during which the work-hardening effect of plastic deformation is balanced by recovery. The creep rate, therefore, remains constant as long as the load and temperature conditions remain the same. The slope of the curve during this stage is the minimum creep rate, i.e. $(\Delta\varepsilon)/(\Delta t)$, and ideally, of course, this should be zero. A component could then be designed with sufficient tolerance to allow for the dimensional changes occurring in stages *I* and *P*. Usually, however, the rate has a finite value and so design for a limited life has to be practised.

T–F — a period of accelerating or tertiary creep leading to eventual fracture. This stage is present either if recovery occurs at a greater rate than work hardening or if the sample begins necking and produces internal cavities. From a service point of view, this stage is only of academic interest since the material should be replaced long before this condition is reached.

The creep curves for any material are sensitive to both stress and temperature. Increasing either has the effect of shortening the secondary creep stage and bringing in tertiary creep more rapidly.

Since it is not feasible to carry out creep tests at every stress and temperature of interest, it is necessary to be able to predict creep behaviour from the results of a few tests. Some sort of relationship between creep rate or creep strain and stress and temperature is therefore needed.

The first systematic investigation into metallic creep was by Andrade who showed that the creep curve for any pure metal under any conditions of stress and temperature could be expressed by a relationship of the type

$$l = l_0(1 + \beta t^{1/3})e^{kt}$$

where l is the length at any time t; l_0 is the length immediately after applying the load; β and k are constants for any given stress and temperature.

Andrade showed that this relationship really describes two types of flow, both occurring simultaneously but to different extents, and being represented by the parameters β and k. Thus, if $k = 0$, $l = l_0(1 + \beta t^{1/3})$ and, by differentiating this with respect to time, the rate of flow $dl/dt = \frac{1}{3}l_0\beta t^{-2/3}$.

This relationship indicates that at long times, dl/dt will approach zero and so refers to a type of flow of decreasing and vanishing rate. This is β or transient flow.

If $\beta = 0$, then $l = l_0 e^{kt}$ and $dl/dt = l_0 k e^{kt} = lk$. Since the creep is only a tiny fraction of l, one can regard l as being constant and so dl/dt is approximately constant also.

This is steady-state creep and it is this type which predominates at high temperatures and stresses. By fitting typical values to β and k, the curves shown in Fig. 8.23 can be plotted. The sum curve is then the total creep curve up to the start of the tertiary stage.

The Andrade relationship indicates the two major types of flow occurring in creep but does not enable one to predict creep behaviour unless the variation of the parameters β and k with stress

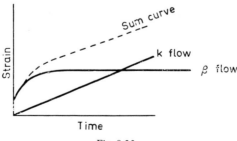

Fig. 8.23

and temperature is known. There are, however, several empirical relationships which enable one to do this, as explained later.

Creep is a thermally activated process involving movement of atoms. A guide to the bonding strength between the atoms in a material is the melting point in absolute temperature Tm. Therefore, similar creep curves are exhibited by different materials at similar fractions of their respective Tm values. Below about $0\cdot25\ Tm$, recovery processes are very slow and so materials will exhibit primary creep but no secondary or tertiary creep. At about $0\cdot5\ Tm$, recovery can occur and secondary creep appears. Above $0\cdot5\ Tm$, tertiary creep may predominate. Temperature is therefore relative. 273 K is a low temperature as regards iron but is sufficient to cause appreciable creep in lead even at low stresses.

Creep mechanisms

An analysis of creep behaviour indicates that creep deformation involves two types of plastic flow:

(*a*) Plastic deformation due to the normal dislocation motion and slip mechanism. This produces work hardening and so the creep rate will either decrease or remain constant depending on the rate of recovery. This sort of deformation predominates in primary and secondary creep.

(*b*) Plastic deformation by viscous flow involving the movement of the grain body as a whole in the grain boundary material which is non-crystalline and so can behave in a viscous manner. This type of flow predominates in the later stages of creep and tends to produce microcracks and voids at grain boundaries. This internal cavitation is one cause of necking down in the tertiary creep stage. Viscous flow deformation is favoured by high temperatures or stresses and, although it occurs predominantly in tertiary creep, it is also going on to some extent during primary and secondary creep.

From a service point of view, the earlier stages of creep deformation are of importance. In these stages, the deformation is mainly by the slip process yet the stresses are relatively low. Some mechanism must therefore be operating which allows dislocation motion to occur at lower than usual stresses if those stresses are maintained for long enough. This mechanism has been shown to involve the diffusion of atoms into vacancies and other point defects in the lattice. Suppose the metal has been placed under stress, has undergone some plastic deformation and has work hardened sufficiently to prevent further deformation under that stress. Theoretically it should not then deform any more. However, at the temperatures operative in creep a dislocation can climb out of its own slip plane, in which it is jammed, into another slip plane, this climb occurring by the diffusion of the atoms in the dislocation into vacancies. This action relieves the pile-up of dislocations on the initial slip plane and so a further increment of slip can occur. Repetition of this climb process then produces the slow, continued deformation particularly evident in secondary creep.

Dislocation climb, at least in pure metals, involves diffusion of atoms of that metal within its own lattice, i.e. self-diffusion, and so it is to be expected that creep is controlled by the same factors which control diffusion. In the creep stages during which recovery is occurring, i.e. chiefly the secondary stage, the creep rate can be expressed as a function of stress and temperature. Thus

$$\text{minimum creep rate } R_0 = \frac{\Delta \varepsilon}{\Delta t} = A_0 \left(\frac{\tau}{\mu} \right)^n e^{-E/RT}$$

where A_0 is a constant; τ is the shear stress acting; μ is the shear modulus; E is the activation energy for creep; n is a constant.

The term $A_0(\tau/\mu)^n$ will be fairly constant for any given conditions of stress and temperature and so a simplified form can be used, i.e.

$$R_0 = \frac{\Delta \varepsilon}{\Delta t} = A e^{-E/RT}$$

This, of course, is an Arrhenius-type relationship (see Chapter 2) and can be placed in the usual log form to give the equation of a straight line.

It is found that the activation energy for creep in pure metals is the same as the activation energy for self-diffusion and this is taken as the basis for describing creep as a dislocation climb process.

The relationship between minimum creep rate and stress at any given temperature can be represented by an equation of type

$$R_0 = \frac{\Delta \varepsilon}{\Delta t} = B \sigma^n$$

where B and n are material constants.

Outside the range of recovery-controlled creep, i.e. at higher stresses or at lower temperatures, the above relationships break down, since dislocation climb is no longer the rate-controlling process. However, they are useful relationships for examining secondary stage creep.

Metallurgical factors influencing creep

There is practically no correlation between the creep properties of a material and its other mechanical properties. This is because creep is so very structure sensitive and is influenced by minor variations in metallurgical condition to a much greater extent than are the other properties. Some of these affecting variables are now discussed.

Grain size

The properties of both the grain body and the grain boundary contribute to creep behaviour, their relative importance being dependent on temperature.

At low temperatures, grain boundaries interfere with dislocation motion by causing pile-up. A fine-grained material, since it contains more boundary material, will tend to resist creep deformation at these low temperatures.

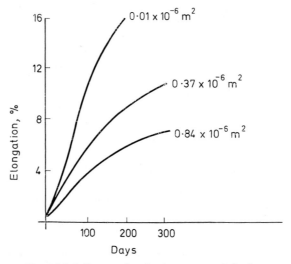

Fig. 8.24. Influence of grain size on creep behaviour

However, at the elevated temperatures which are of more interest in creep, boundaries are capable of acting both as sinks for atoms and sources of vacancies. A fine-grained material would therefore encourage dislocation climb and so increase the amount of creep due to the slip process. Fine grains will also provide more opportunity for viscous sliding.

The temperature of inversion of the role of grain boundaries is termed the equicohesive temperature and this, as one would expect, is related to the melting point of the material.

The beneficial effects of a coarse grain size are illustrated in Fig. 8.24. This actually refers to lead loaded at 3·5 MN/m² and 293 K, but the pattern would be similar for other materials.

Prior strain

Previous cold work can modify creep behaviour in the primary and secondary stages, since in a cold-worked material the slip process is less easy. Pre-straining may therefore be beneficial, as is illustrated in Fig. 8.25 which also refers to lead. Obviously, the operating temperature must not be capable of triggering off recrystallisation.

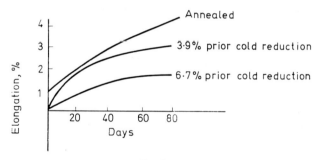

Fig. 8.25

The influence of prior cold work is not progressive and it is usually found that there is an optimum amount, pre-strains greater than this amount actually increasing the creep rate. This is illustrated in Fig. 8.26, which refers to a stainless steel.

The locus of the maxima indicates that the optimum amount of prior cold work is lower, the higher the temperature of operation. Prior strain above the optimum triggers off recrystallisation and rapid softening.

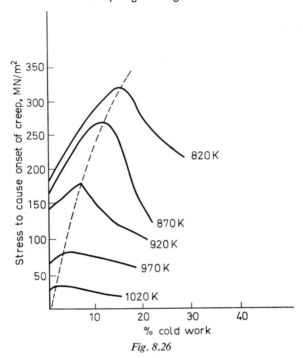

Fig. 8.26

Formation of solid solutions

Substitutional solid solution formation will usually improve creep resistance since it involves the production of lattice strains which hinder dislocation movement. Interstitial solid solution formation, such as occurs in BCC materials, may also give better creep resistance because of the Cottrell locking of dislocations.

The frictional resistance offered by solute atoms to the movement of dislocations is useful at moderate temperatures in steels (up to 720 K). Many alloys, however, operate above about 900 K—the superalloys—and at such temperatures oxidation resistance becomes important. For this reason, many of the superalloys contain large quantities of chromium. Since metals such as nickel and cobalt have better creep resistance than iron, these superalloys tend to be based on Ni–Cr–Fe–Co solid solutions rather than being steels. They are required to operate at up to 0·75 *Tm*, and at such temperatures diffusion rates are high and so solute atoms offer very little frictional resistance to dislocation motion. At high operating temperatures, therefore, the contribution of solid solution strengthening to creep resistance is not very great.

The amount of solute in solution must also be limited, since as more solute is added the solidus temperature is lowered. Hence, the atomic mobility in a highly alloyed solution at a given temperature may outweigh the benefits of lattice distortion, since such an alloy may be very close to its solidus temperature.

Complex solid solutions may therefore be beneficial for low operating temperatures but, in the range above $0.5\ Tm$, some more stable strengthening agency is needed, i.e. some form of precipitation strengthening.

Precipitation or dispersion hardening

Most commercial creep-resistant alloys are based on complex solid solutions but contain a very finely dispersed precipitate. If this precipitated phase is to be effective in retarding creep, it must remain finely enough dispersed to interfere with dislocation motion. The following are three main ways in which such dispersion is ensured.

(*a*) Many precipitated phases grow rapidly at creep temperatures and so would soon become useless. It is, however, possible to adjust the composition of the alloy so that a succession of phases is precipitated one after the other, so maintaining the strengthening effect. Some ferritic creep-resistant steels containing chromium and molybdenum behave in this way. Obviously, such alloys have a finite operating life which ends when the supply of precipitate becomes exhausted.

(*b*) The driving force for any precipitation process is the interfacial energy between particle and matrix. If this is high, precipitation is more likely to occur. If, therefore, a cluster can be produced which is isomorphous or very nearly isomorphous with the parent lattice, then there will be less tendency for precipitation and growth. Such coherent clusters or precipitates are responsible for much of the creep resistance of the nickel-base alloys.

(*c*) For two particles of precipitate to grow into one, atoms must be transported through the lattice i.e. atoms from one particle must redissolve in the matrix, travel through the lattice and deposit on the other particle. This sort of transport and growth can be retarded if the precipitate is highly insoluble in the matrix. Sintered alumina powder (SAP) and internally oxidised materials are typical here and make use of the low solubility of metal oxide in metal.

However the precipitated particles are produced, i.e. by either age hardening or dispersion hardening, the spacing of the particles is critical. The particles must be of sufficient fineness and sufficiently finely dispersed to prevent dislocation motion. The critical nature of this spacing has been discussed already in Chapter 7.

Selection of metals for creep service

Selection must obviously be on the basis of actual testing but some general principles can be outlined.

(1) The service temperature should not approach the prevailing recrystallisation temperature since any such microstructural change will accelerate the creep rate. This leads to a division of metals into three general groups. Lead, tin, zinc and their alloys recrystallise near to 273 K and so will creep under very low stresses at normal temperatures. Lead, for example, will creep at room temperature even under its own weight.

Aluminium, magnesium, copper and their alloys recrystallise in the range 370–520 K and so are capable of resisting creep at normal temperatures under low stresses,

Iron, nickel, tungsten, chromium, molybdenum and their alloys have high recrystallisation temperatures and these form the basis of the usual alloys for high temperature use, e.g. stainless steels, nimonics, inconels, chromium base alloys.

(2) Polymorphic changes should be absent over the operating temperature range since these would produce variation in creep characteristics. For this reason, creep-resistant steels are stabilised either as ferritic or as austenitic.

(3) Solid solution formation produces lattice distortion and, if diffusion rates are low, hinder dislocation motion. The solid solution hardening effect is, of course, greatest where the alloying atoms differ widely in size from the parent atoms, but this idea cannot be pursued to extremes since solubility usually decreases as size difference increases. However, a complex solid solution of high melting point elements is often the basis of a creep-resistant alloy and this is further strengthened by precipitation treatments.

(4) The material should not be put into creep service while it contains residual stresses. It should be of relatively coarse grain size and, partly for this reason, castings are often preferred.

(5) If the service temperature is too low to cause recrystallisation, prior cold work may be beneficial.

(6) Creep-resistant alloys operate at elevated temperatures and often in atmospheres containing the products of combustion. It is, therefore, usual to find that a creep-resistant alloy is also a corrosion- and oxidation-resistant alloy.

(7) The material must be structure-stable at operating temperature, as far as is possible. Hence, microstructural changes such as continuation of tempering, over-ageing, grain growth, etc., must be absent and this often means carrying out any prior heat treatment at least to the proposed service temperature.

CORROSION AND OXIDATION OF METALLIC MATERIALS

Corrosion involves a direct loss of metal running into many millions of pounds annually. There is also indirect wastage of resources and material involved in painting, protection and in the deliberate overdesigning practised to give reasonable service life. It is also as well to remember that corrosion can accelerate failure due to fatigue or brittle fracture.

Corrosion can be defined as the changing of a metal from its elemental form into a chemically combined form. Such a definition also covers the process of oxidation. Any metal, unless it is a 'noble' metal such as gold, silver, etc., will always tend to revert back to the combined state since this is the state of lowest energy. This is the reason why most metals are found in nature as chemical compounds (ores) and why energy must be supplied to extract the metals from these ores.

It is found that corrosion is electrochemical in nature and is accompanied by the passage of an electric current. This means that ions and electrons undergo transfer during the process. This energy transfer can be illustrated for oxidation (dry corrosion) and for wet corrosion.

Oxidation processes

At the surface of a solid metal the regular lattice pattern comes to an abrupt end. The surface ions tend to be more reactive than the ions within the body because of the inequality of bonding on the surface. Surface ions will therefore readily establish bonds or links with external media such as oxygen, even at normal temperatures. This tendency to form surface films is so pronounced that it is impossible, except in high vacuum, to prepare a perfectly bare metallic surface. The surface will normally be covered with a thin film of its own oxide and this film is often self-healing.

Oxidation goes forward in a number of distinct stages. Oxygen molecules coming into contact with the metal surface dissociate into separate oxygen atoms which are then adsorbed on to the surface, probably by van der Waals' forces, as show in Fig. 8.27(a).

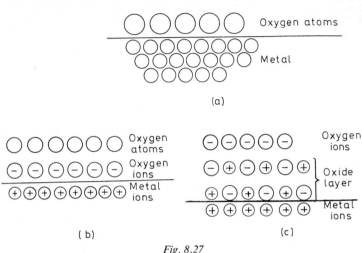

Fig. 8.27

The oxygen atom has an electron configuration of 2–6 and, if electrons are available, will tend to pick up electrons to produce the stable octet arrangement 2–8. Free electrons are available in the metal and these migrate into the vacant energy levels in the oxygen atoms, so converting them to O^{2-} ions. An ionic bond is now established between the positive metal ions and negative oxygen ions (Fig. 8.27(b)).

This layer of oxygen ions builds up and, to maintain charge balance, the metal and oxygen ions arrange themselves so that each type of ion is surrounded by ions of the opposite type. Such an arrangement obviously fulfils the definition of crystallinity and, in fact, the metal oxide is an ionically bonded crystalline solid (Fig. 8.27(c)).

The growth of the oxide films goes on by interdiffusion of metal and oxygen ions but, as the oxide layer thickens, the attractive forces between metal and oxygen ions decreases and growth slows down. In some cases, it may cease altogether. On aluminium, for example, the Al_2O_3 film reaches a limiting thickness of about 20×10^{-10} m at 293 K.

In cases where the oxide film is coherent with the underlying metal lattice, growth often stops when the limiting thickness is reached. The film is usually tight and impervious and offers protection against further oxidation and corrosion, i.e. the metal is passivated. Al_2O_3 on aluminium and Cr_2O_3 on chromium are typical of such films.

In most cases the oxide crystal structure is not coherent with the metal structure and so strains are developed at the metal–oxide interface which lead to cracking of the film. Further metal is thus continuously exposed to oxygen and so such films are non-protective. Porous oxide films are typically produced on the alkali and alkaline earth metals. The oxide film in such cases has a smaller volume than the metal from which it was formed. One can imagine that this reduction in volume of the surface layers produced when an oxide is formed will place the oxide layer in tension and will produce cracking.

More impervious oxides are produced on most other metals and these oxides tend to be in compression, since the volume of the oxide is greater than the volume of metal consumed in its formation.

Generally, oxides have higher melting points than the metals from which they are formed and so may give some protection against further attack. In the case of molybdenum, however, the oxide is volatile above about 820 K and oxidation becomes catastrophic as bare metal is continuously exposed. This is one of the factors which hampers the application of molybdenum alloys as high-temperature materials.

The growth of an oxide film can occur in a number of different ways, each way being capable of being described by a 'growth law'. The 'parabolic growth' law is observed with those metals in which the growth of the oxide film is completely governed by the rate of diffusion of metal and oxygen ions. At any given temperature, the increase in the film thickness F with time t is inversely proportional to the thickness F, i.e.

$$\frac{\mathrm{d}F}{\mathrm{d}t} = k \cdot \frac{1}{F}$$

where k is a constant for the metal related to its diffusion coefficient. On integration, $F^2 = kt$. At room temperature, metals such as iron, nickel and copper oxidise in this parabolic manner.

The 'linear growth' law is observed when the rate of oxidation at any given temperature is constant, i.e. it is not affected particularly by diffusion rates. Then $F = xt$, where x is some constant. Metals such as sodium, potassium behave in this manner, since the oxide films are porous and allow continuous access of oxygen to the metal surface.

20*

Metals such as aluminium, chromium and zinc are metals on which the oxide film stops growing after a critical limiting thickness is reached. These are examples of the operation of the not very common 'logarithmic growth' law $F = k_1 \log(yt+1)$, where k_1 and y are material constants for any given temperature. Such oxide films are quite protective. The cessation of the growth of the oxide film in such cases is often the result of high electrical resistance of the oxide. This may be due to ionic resistance or to resistance to the passage of electrons but, in either case, oxidation will eventually stop, since metal ions become incapable of contacting oxygen ions and vice versa.

These laws of oxide growth refer to constant temperature. In those oxidation processes which involve diffusion, it is to be expected that the oxidation rate will be temperature sensitive. This sensitivity can, as usual, be expressed using an Arrhenius relationship:

$$\text{oxidation rate} = \frac{dF}{dt} = Ae^{-E/RT}$$

and the plot of log rate against $(1/T)$ should give a straight line which should be capable of extrapolation. However, the relationship is only useful as a general guide, since the oxidation process may change as temperature changes. For example, magnesium oxide films thicken in a parabolic manner up to about 770 K but then the film becomes porous and cracked and further growth more closely follows the linear law. The Arrhenius relationship will not then be applicable. Because of the porous nature of most oxide films, it is usually necessary to remove them completely before applying any protective coating to the metal. In wet corrosion processes, for instance, the presence of the oxide may actually accelerate the attack on the underlying metal.

Wet corrosion processes

Since corrosion involves the passage of an electric current from one part of a metallic system to another, then (*a*) some agency must be present to initiate the attack and also carry the current—the *electrolyte;* (*b*) some agency must be maintaining a difference in voltage.

By definition, an electrolyte is any fluid which is capable of carrying a current, and for our purpose this usually means either aqueous solutions of inorganic compounds or molten inorganic compounds. Typical compounds are chlorides, sulphates, nitrates of the metals. Natural waters contain dissolved salts and these are really dilute solutions of electrolytes. Such solutions are capable of carrying a current since they are ionised solutions (see Chapter 1). Inorganic materials which contain the ionic bond will dissociate into their respective ions on solution or melting, each ion carrying a charge equivalent to its valency. Thus sodium chloride NaCl dissociates into Na^+ and Cl^-, while ferric chloride $FeCl_3$ dissociates into Fe^{3+} and $3 Cl^-$ ions. The number of positive charges at any instant balances the number of negative charges and so the solution as a whole is electrically neutral. However, if charged electrodes are placed in the solution, the positive ions move towards the negative pole and vice versa, and this movement of charges constitutes conductance.

It must also be remembered that bulk metal is a collection of positive metal ions embedded in a common valency electron cloud of negative energy.

The behaviour of a metal in an electrolyte can now be considered.

Solution potential

If a metal is placed in an ionised solution, there is a tendency for metal ions to leave the metal and enter the solution. Quite simply, one can consider this as an attempt to even out the concentration of positive ions between the metal which is rich in ions and the solution which is less rich in ions. The loss of positive metal ions from the metal will leave the metal with an unbalanced negative charge due to the excess electrons. Eventually, the electrostatic attraction between the positive metal ions in the solution and the residual negative charges on the metal produces a state of equilibrium, as illustrated in Fig. 8.28. This equilibrium potential on the metal is called Solution Potential.

The bar of zinc in Fig. 8.28 is acting as an *anode* since an anode for our purpose is considered as that part of a metal system at which current leaves the metal and enters the solution. The current in this context is considered as a flow of positive charges. Such a flow of positive metal ions into the solution constitutes dissolving of the metal and so, in a corrosion process, it is always the anode which is attacked.

The solution potential has different values with different metals and different conditions of temperature, electrolyte concentration, etc. It can also vary with the metallurgical condition of a metal. Obviously, if the excess negative potential on the bar of zinc in Fig. 8.28 were to be drained off in some way, then the equilibrium

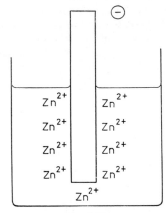

Fig. 8.28

would be disturbed and more Zn^{2+} ions would pass into solution in an effort to restore the equilibrium. This sort of situation arises in a corrosion cell, a model of which is illustrated in Fig. 8.29.

In this cell, it is assumed that material A has the highest solution potential and so material C could be thought of as being positive towards A. A is therefore regarded as the anode.

Material A will lose A^+ ions into the solution, as indicated, and will assume a negative charge. However, A is in electrical contact with C and since C has the lower solution potential, electrons will drain from A to C. This disturbs the equilibrium on A and so

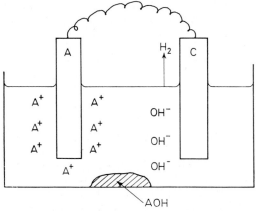

Fig. 8.29

more A^+ ions will pass into solution, i.e. the anode A continues to corrode or dissolve away as long as electrons pass from A to C.

Now let us assume that the electrolyte is ordinary water. This ionises to some extent to give hydrogen ions H^+ and hydroxyl ions OH^-

$$H_2O \rightarrow H^+ + OH^-$$

The solution around C will therefore contain a few C^+ ions, because of the solution potential effect, and will also contain H^+ and OH^- ions. Negative electrons draining on to C from A are neutralised as they contact the H^+ ions.

$$H^+ + e^- = H$$
$$H^+ + e^- = H$$
$$\Big\} H_2$$

and so hydrogen gas bubbles may be evolved at C. Pole C is therefore the pole at which positive charges leave the solution and enter the metal; such a pole is referred to as a cathode. It is not subject to corrosive attack.

Since H^+ ions are being removed from around the cathode as hydrogen gas, the solution here becomes enriched in OH^- ions. In time, A^+ ions from around A will meet OH^- ions from C and a chemical reaction may occur as charges are neutralised:

$$A^+ + OH^- = AOH$$

This reaction produces a metal hydroxide, the rust or corrosion product. Metal hydroxides are usually insoluble and so the corrosion product is often precipitated. This, of course, effectively removes ions from solution and so the corrosion process goes on. In ferrous materials, the reactions would be

$$Fe^{2+} + 2(OH)^- = Fe(OH)_2$$

The green precipitate of ferrous hydroxide $Fe(OH)_2$ is then gradually converted to the familiar red rust by oxidation from the air:

$$4\,Fe(OH)_2 + 2H_2O + O_2 = 4\,Fe(OH)_3$$

The actual point of deposition of the precipitate will depend on the relative diffusion rates of A^+ and OH^- ions. In some cases, it may actually be deposited on the cathode surfaces where it forms a non-conducting film and so can stifle further corrosion. Usually, however, the corrosion product is deposited away from both anode and cathode and so has no protective action.

The bubbles of hydrogen gas liberated at the cathode may also slow down the rate of attack if they cling to the cathode and so insulate the cathode from the solution. However, this 'polarisation' action is nullified if the solution contains dissolved oxygen, since this combines with the hydrogen to re-form the water molecule $H^+ + H^+ + O^{2-} = H_2O$. The protective hydrogen gas film is thus removed. Industrial boiler waters are often de-aerated by vacuum or by chemical treatment in order to promote the protective effect of polarisation. Steel will not corrode in waters which contain no dissolved oxygen.

Origins of corrosion currents

Contacts between dissimilar materials in the presence of an electrolyte

A metal in an ionised solution loses its own ions into that solution and assumes a solution potential or voltage. The value of this solution potential is therefore a reflection of the ease with which the metal will corrode and so a knowledge of this value would be useful. However, in order to measure this voltage, it is necessary to complete the circuit with a second electrode and since this second electrode will have its own value of solution potential, all that can be measured is the difference in potential. It is therefore necessary to choose a standard or reference electrode, give this an arbitrary solution potential value of zero, and then compare other materials against this. Since solution potential is also influenced by factors such as temperature, concentration of electrolyte, etc., it is also necessary to standardise the test conditions.

The standard reference electrode is actually hydrogen gas, made to act as a continuous conducting film by bubbling it over a plate of activated platinum at 1.0133×10^5 N/m^2 pressure. The standard electrode is contained in an electrolyte consisting of molar solution of hydrochloric acid at 298 K and is connected to the test metal which is itself immersed in a molar solution of its own ions at 298 K.

The solution potential measured in this way is referred to as the standard electrode potential E^0. Some typical values are given in Table 8.7. Some general points emerge from a study of this table:

(*a*) The higher the negative value of E^0 the more easily is the material corroded. Aluminium and chromium appear at the base end of the series but the relevant E^0 values refer to the bare metal surfaces. In practice, these metals are covered with a

Table 8.7

	Electrode	E^0 V	Electrode	E^0 V	
'Base' end	Li–Li$^+$	-3.02	Ni–Ni^{2+}	-0.235	
	K–K$^+$	-2.92	Sn–Sn^{2+}	-0.14	
	Ca–Ca^{2+}	-2.87	Pb–Pb^{2+}	-0.125	
	Na–Na$^+$	-2.71	H$_2$–2H$^+$	0	
	Al–Al^{3+}	-1.69	Sb–Sb^{3+}	0.203	
	Mn–Mn^{2+}	-1.1	Cu–Cu^{2+}	0.34	
	Zn–Zn^{2+}	-0.76	Ag–Ag$^+$	0.79	
	Cr–Cr^{3+}	-0.65	Hg–Hg$^+$	0.80	'Noble' end
	Fe–Fe^{2+}	-0.44	Au–Au^{3+}	1.35	
	Cd–Cd^{2+}	-0.40	Cl$_2$–2Cl$^-$	1.36	
	Co–Co^{2+}	-0.28			

protective film of their own oxides and so do not corrode. It is, however, useful to realise that if the Al$_2$O$_3$ film is removed from aluminium alloys or if the Cr$_2$O$_3$ film is removed from, say, stainless steels, then such materials are no longer corrosion resistant. Fortunately, the films are self-healing and so film removal would have to be continuous.

(b) If two materials are brought into electrical contact in the presence of an electrolyte, the anode will be the one with the highest negative value of E^0 and this is the member of the pair which will be corroded.

(c) The rate of attack on the anode in a dissimilar metal cell will increase as the algebraic difference between the E^0 values increases.

(d) Anodic materials are capable of displacing more cathodic metals out of solution. For example, zinc dipped into copper sulphate solution will be dissolved but, at the same time, a precipitate of metallic copper is produced, i.e.

$$Zn + Cu^{2+}SO_4^{2-} = Zn^{2+}SO_4^{2-} + Cu$$

In practice, dissimilar materials often come into contact and, quite often, severe corrosion of the anodic member does occur. However, before assessing the potential danger of any such contacts, the relative areas of anode and cathode need to be considered. What must always be avoided is a large cathode area in contact with a small anode since, in such a case, all the corrosive attack is concentrated on to the small anode and pitting corrosion results. For example, copper plate fastened with steel bolts would be a dangerous combination, whereas steel plate fastened with copper bolts would not—the steel would still corrode but the attack would now be evenly spread over the large anode area.

The E^0 series is useful as a preliminary guide to corrosion behaviour but the values really only refer to the standard test conditions. More practical series are available (galvanic series) which include alloys as well as metals and refer to normal conditions using common electrolytes such as sea- or rain-water.

Some typical examples of corrosion as a result of contacts between dissimilar materials can now be considered.

Heterogeneity of composition or microstructure can generate different E^0 values even within the same piece of material. Austenitic stainless steel (18Cr : 8Ni : 0·1C) can be prone to failure by weld decay after welding—a popular method of fabricating this material. This type of steel should be a single-phase solid solution of $Cr + Ni + C$ in austenite and its corrosion resistance is due to the presence of a coherent film of Cr_2O_3 on its surface. For this oxide film to be thick enough to be protective, something over 11% of chromium must be present in solid solution. Chromium, however, is a powerful carbide-forming element and if the steel is heated for any length of time in the region 800–1100 K, the chromium and carbon in the steel are likely to combine and give a precipitate of chromium carbide on the grain boundaries, hence removing chromium from solid solution. Such a temperature range is present at the back of a weld because of conduction of heat into the parent metal. A situation such as that illustrated in Fig. 8.30 could arise.

The grain body A contains 18Cr : 8Ni : 0·1C in solid solution. The grain boundary carbide precipitate B is rich in Cr and so there will be a zone C on the grain edge which has been denuded of solid solution Cr and may contain less Cr than is required to maintain

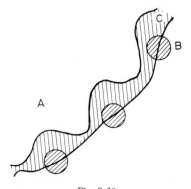

Fig. 8.30

corrosion resistance. There are therefore two dissimilar compositions in contact, and in the presence of an electrolyte corrosive attack and cracking will penetrate via zone *C*. This sort of embrittlement can be prevented by heating after welding to a high enough temperature to cause re-solution of the chromium carbides (over 1200 K), followed by rapid cooling to suppress further precipitation. This, of course, is only useful if the whole of the welded component can be so treated. The difficulty is best overcome by confining welding fabrication to the stabilised grades of stainless steel. These contain small quantities of elements such as titanium or niobium which are even more powerful carbide-forming elements than is chromium. Any carbides which are precipitated will then be titanium or niobium carbides and not chromium carbides and so chromium is not stripped from solid solution.

A similar sort of intercrystalline embrittlement can arise in duralumin-type alloys if these are over-aged. Over-ageing causes visible precipitation of $CuAl_2$, often on the grain boundaries, and this means that copper is being removed from solid solution in the aluminium. The $CuAl_2$ precipitate is rich in copper, while the grain body is rich in aluminium and so, again, two dissimilar materials are in contact and intercrystalline corrosion may occur in the presence of an electrolyte.

The dissimilar materials in contact need not both be metals. A metal oxide or paint film, for instance, is usually cathodic to the metal underneath and one can visualise a situation in which a break in an oxide film becomes covered with water, as in Fig. 8.31.

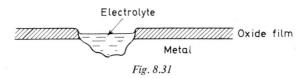

Fig. 8.31

This constitutes the particularly dangerous combination of a small anode in contact with a large cathode. All the attack is concentrated on the small anode and pitting corrosion results. Such a pit constitutes a stress concentration and may trigger off fatigue failure or brittle fracture. From this point of view, a partial surface coating is more dangerous than no coating at all.

There are, of course, many cases where contacts between dissimilar materials are deliberately produced—brazed and soldered joints, hard-faced components, bearing components, and the like. It is as well to remember that all such components are prone to differential attack under corrosive conditions.

Differences in ion concentration

The solution potential effect involves the passage of metal ions into the solution. If the solution already contains a large quantity of ions, then fewer metal ions will pass into solution and so the solution potential will be lower than usual, i.e. the metal becomes more cathodic. The reverse is also true. If, therefore, a piece of metal is in contact with an electrolyte which varies in concentration, those areas of metal in contact with dilute solution will become anodic to those areas in contact with more concentrated solution and corrosion will occur.

Natural waters contain dissolved solids and the concentration may vary with depth and degree of stagnation. In a stagnant pool, for example, the surface layers would be less concentrated than deeper layers, simply due to the effects of rain. Steel piling in such a pool will often be preferentially corroded near the water surface. Differences in velocity of a moving electrolyte can also have similar effects. For example, a propellor spinning in sea-water has a high peripheral speed and a relatively low shaft speed. The water near the shaft is relatively little disturbed and ions can concentrate here. As a result, the periphery becomes anodic and is preferentially attacked. A similar situation arises with liquids travelling at varying velocities in pipes.

Differences in oxygen concentration

The natural oxide film on a metal can only be maintained if it is in contact with air or oxygen. Shielding the metal from air can cause film breakdown and it is found that those areas of metal in contact with a poor oxygen supply become anodic to those areas which are well aerated. A very simple example of this is illustrated in Fig. 8.32.

Because of the shape of the water droplet, dissolved oxygen can reach the peripheral regions *B* more easily than it can penetrate to the metal under *A*. As a result, the periphery becomes cathodic

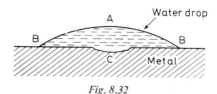

Fig. 8.32

and pitting corrosion develops at C. This sort of pock-marking is often seen on steelwork exposed to salt spray.

The same sort of effect can occur whenever a metal surface is partly covered and shielded from air. Even the protective Cr_2O_3 film on stainless steel will break down if shielded from air and cases have been reported of rapid rusting of stainless steel underneath plastic washers used in bolting.

Stress corrosion cells

Plastically deformed metal is in a higher energy condition than unstrained metal and is more prone to chemical attack. Anode–cathode regions may therefore be set up in a metal which has been subject to differential amounts of work hardening. Caustic cracking around rivet holes in boiler shells is a typical example. In driving a rivet, the metal underneath the head is work hardened. Slight seepage of water often occurs near a riveted joint and, usually, the boiler water will have been treated with caustic salts for softening. As the seeping water evaporates, it deposits its salt content and eventually the solution under the rivet head may become quite concentrated. Preferential attack of the work-hardened metal can then occur leading to eventual cracking.

Grain boundary material in metals work hardens more rapidly than the grain body material and so, in a severely cold-worked component, the boundaries could become anodic to the grain bodies. Preferential attack via the boundaries can give rise to intercrystalline embrittlement, e.g. season cracking in cold worked brass.

Differential corrosion due to residual stress in fusion welds and corrosion fatigue are also examples of stress corrosion. Stress relief annealing (recovery) will reduce the risk of this sort of corrosion.

Thermodynamic approach to corrosion—Pourbaix diagrams

The table of E^0 values given previously gives some idea of the corrosion resistance of metals but it refers only to the effects of anode–cathode potential difference on the corrosion process.

Many corrosion processes depend also on the acidity of the environment as well as on anode–cathode differences. For example, a corrosion reaction of the type $Zn \rightarrow Zn^{2+} + 2e$ is purely electrochemical and depends on the E^0 value. However, a reaction

$$Zn + 2H_2O \rightarrow Zn(OH)_2 + 2H^+ + 2e$$

involves both the production of electrons e and of hydrogen ions H^+, and hence depends both on potential difference and on the initial hydrogen ion concentration (acidity) of the solution.

The hydrogen ion concentration of an aqueous solution is measured by its pH value, which is the negative power of the hydrogen ion concentration in g ions/l.

Pure water ionises slightly, $H_2O \rightharpoonup H^+ + OH^-$, and actually contains 10^{-7} g ions/l of H^+ at 298 K. This, therefore, has pH = 7 and is regarded as a neutral solution.

Acid solutions are richer in H^+ ions and so have pH values below 7. Alkaline solutions are rich in OH^- ions (contain few H^+ ions) and so have pH values greater than 7. The pH scale extends from pH = 0 to pH = 14 and it should be remembered that it is really a log scale. Hence, an acid solution of pH = 3 contains ten times more H^+ ions than a solution of pH = 4.

Suppose zinc is placed in ordinary water. Because H^+ and OH^- ions are present, the water is an electrolyte and so the usual solution potential effect will occur, i.e. $Zn \rightarrow Zn^{2+} + 2e$. The electrons liberated would pass over to a cathode if this were available and would be neutralised as usual by the H^+ ions near the cathode, so causing evolution of hydrogen gas:

$$2H^+ + 2e = H_2$$

Provision of a greater quantity of H^+ ions near the cathode (reducing the pH of the solution) will obviously increase the rate of the reaction $2H^+ + 2e = H_2$ and, hence, increasing the acidity of the solution will accelerate the corrosion of the zinc. All metals having E^0 more negative than hydrogen (base metals) will behave in this way since they are anodic to hydrogen.

The noble metals are, however, cathodic to hydrogen and so the above reaction will not occur, i.e. hydrogen gas cannot be a product at the cathode. These metals may, however, corrode if oxygen is dissolved in the electrolyte since then another reaction becomes possible, i.e.

$$\tfrac{1}{2}O_2 + 2H^+ + 2e = H_2O$$

i.e. the electrons coming from the anode can be neutralised by H^+ in the presence of oxygen and so corrosion of the anode can proceed. The reaction

$$\tfrac{1}{2}O_2 + 2H^+ + 2e = H_2O$$

has an E^0 value of 1·23 V and so any metal with an E^0 value below this figure could be corroded by this mechanism. As before, the rate of corrosion at the anode will be a function of the pH of the solution since H^+ ions are evolved.

The Pourbaix diagram attempts to account for the influence of both potential differences and pH on the corrosion of metals. The diagrams are calculated from the thermodynamics of the reactions involved and refer to hydrogen gas being evolved at standard pressure or O_2 gas being available at a standard pressure. They are available for metal–water systems and are illustrated by the following examples.

$Zn–H_2O$ *system*

If zinc is in contact with water, the following reactions are possible.

(a) $Zn \rightarrow Zn^{2+} + 2\,e$—solution potential effect; zinc ions pass into solution.

(b) $Zn + 2\,H_2O \rightleftharpoons Zn(OH)_2 + 2\,H^+ + 2e$—precipitation of zinc hydroxide.

(c) $Zn(OH)_2 + 2\,H^+ \rightleftharpoons Zn^{2+} + 2\,H_2O$—re-solution of zinc hydroxide in a solution of low pH value.

(d) $Zn(OH)_2 \rightleftharpoons ZnO_2^{2-} + 2\,H^+$—re-solution of zinc hydroxide to give zincate ions in solutions of high pH value.

(e) $Zn + 2\,H_2O \rightleftharpoons ZnO_2^{2-} + 4\,H^+ + 2\,e$—direct production of the zincate ion in solutions of high pH value.

Reactions (c) and (d) are purely chemical and depend on the pH of the solution only. For example, in acid solution, reaction (c) would occur while (d) would occur in alkaline solutions. Reactions (a), (b), (e) are electrochemical in nature, since these involve the transfer of electrons. Reaction (a) is influenced by potential difference only, whereas (b), (e) are influenced by both potential difference and pH value, since they involve both electron transfer and H^+ concentration.

Reactions which are influenced by voltage only will appear as horizontal lines on a E^0/pH plot, reactions influenced by pH only appear as vertical lines, and reactions governed by both pH and voltage appear as sloping lines.

Fig. 8.33 illustrates the Pourbaix diagram for the $Zn–H_2O$ system. Both the production of Zn^{2+} ions and of ZnO_2^{2-} ions represent the dissolving of zinc into solution and so these areas indicate conditions under which corrosion is possible.

Zinc hydroxide $Zn(OH)_2$ is almost insoluble under certain conditions of pH value and so may produce a protective deposit over the surface of the metal. This area of the diagram, therefore, indicates conditions of possible passivity. The other domain on the

an upper limit to the alkalinity of the solution since, as the diagram shows, it is possible to produce more corrosion in highly alkaline waters at low potentials by the formation of soluble ferrites.

Pourbaix diagrams are obviously applicable to predicting means of corrosion prevention but they have certain limitations. They give no indication as to the probable rate of a corrosion reaction and the 'passivity' regions need to be interpreted carefully since the precipitates of metal hydroxide do not always provide a protective film.

Protection against corrosion

Protective surface films

(1) *Chemical treatments*—The aim here is to alter the composition of the surface layers and to produce a metal compound which is more noble or cathodic than the metal itself. *Phosphate coatings* can be applied to ferrous materials and zinc alloys by pickling in a hot solution of phosphoric acid. The metal phosphate layer produced is reasonably protective of its own accord but is often used as a basis for subsequent painting or enamelling. The process is operated under names such as parkerising, bonderising, granodising. *Chromate* surfaces are also reasonably protective on ferrous materials, zinc and light alloys. Potassium and sodium chromates are powerful oxidising agents and the protective action on steel and aluminium alloys is probably due to the production of a protective oxide film rather than a chromate film. The natural protective film on aluminium alloys can also be thickened by *anodising*. The alloy is made the anode in a hot bath of chromic acid with steel as a cathode. The Al_2O_3 film is thickened to about $2 \cdot 5 \times 10^{-6}$ m and is matt enough to be subsequently oiled, dyed or enamelled.

(2) *Metal coatings*—Hot-dipped coatings may be applied, usually to ferrous materials. The clean steel is immersed in molten zinc, tin, lead or aluminium. The films usually consist of an outer layer of protective metal bonded to the base metal via an alloy layer. The behaviour of a metal-coated component under corrosive conditions depends to some extent on the relative E^0 values of the metals. For example, with galvanised iron the zinc is anodic to the steel and, if the coating is ruptured so that an electrolyte can penetrate, the zinc will be corroded in preference to the steel. This is an example of sacrificial cathodic protection. On the other hand, it would be the steel which would corrode at a break in a tin coating, since iron is anodic to tin. This would also give the combination of large cathode with small anode, resulting in pitting corrosion.

Metal coatings can also be produced by *electrodeposition* of a very wide range of both pure metals and alloys. The article to be plated is made the cathode in a suitable electrolyte with the protective material as the anode. The coatings are thin (2.5×10^{-7} to 2.5×10^{-5} m) and there is very little alloy layer bonding with the base metal. Because of the random way in which the protective ions are deposited, the coatings tend to be highly strained and rather brittle and also somewhat porous. The inherent porosity requires the use of multi-coatings. For example, good-quality chromium plating on steel would follow the sequence Fe–Cu–Ni–Cr. The deposited materials are usually the more cathodic metals and their alloys, e.g. Cr, Cd, Cu, Au, Pb, Ni, Pt, and the process is applicable to nearly all metallic materials and to non-metals if these are first of all made conductive with a graphite coating. *Electrodeposition* of tin coatings on steel for canning purposes has practically replaced hot-dip tinning.

Protective surface coatings can also be produced by *cementation*, a process whereby the component to be protected is heated in contact with a solid powder of the protective metal. The coating is produced by inward diffusion of the protective metal and so an alloy layer bond is produced. Specific examples are sherardising, used for zinc coating of steel, calorising, for aluminium coating, and chromising.

Metal spraying is often used to give protection *in situ* to large structures such as bridges. The protective metal is melted in an electric arc or gas jet and then projected as a fine mist of droplets by compressed gas. The molten droplets chill on to the surface to be protected, which has already been roughened by sand blasting, and little alloy layer formation occurs. The coatings are rather soft and porous but the method is widely used for the protection of structural steelwork with zinc or aluminium.

Metal coatings can also be produced by *cladding*. A sandwich of base metal sheet and protective metal sheet is rolled or pressed at a temperature high enough to cause pressure welding. The process is used to bond pure aluminium surfaces on to aluminium alloys (alclad) and stainless steel on to mild steel.

(3) *Use of paints and varnishes*—Paints are often composed of a metal oxide or chromate suspended in a carrier such as a drying oil. These paints set or harden by the oxidation of the drying oil. A typical drying oil is linseed oil. This is actually a polymer which cross-links and so hardens on oxidation. The setting action involves a volume change which produces porosity in the film and so multi-coats are needed. The protectives in the paint are oxides of lead, iron, zinc, titanium or chromates of lead. These are cathodic to the

basis metal and are also oxidising in nature so producing some natural passivity. Plastic paints are composed of polymers such as PVC, polyethylene, rubber, etc., dissolved in solvent. They set by evaporation of solvent and this again leads to porosity and the necessity for multi-coatings.

Tar and bitumen coatings are often applied to underground metalwork. These are applied molten and set by solidification. As such, they are not porous and offer good protection.

(4) *Vitreous enamels*—Enamels are really amorphous ceramics, i.e. glasses. A mixture of a metal oxide and siliceous material is sprayed over the surface of the metal to be protected and then the temperature is raised to cause the oxide and silica to react to form a fusible glass. It is this glassy film which gives the almost total resistance to corrosion. Unfortunately, the film is brittle and easily broken.

Use of inhibitors

An inhibitor is a substance which reduces the rate of corrosion. It does not actually prevent corrosion and so protection is temporary. Inhibitors are widely used in the treatment of boiler waters, in acid pickling of steel and to give protection during transit or storage.

Soluble inhibitors may be added to boiler waters to produce stifling precipitates either on potential anodes or cathodes. Typical examples are phosphates, dichromates, silicates, hydroxides and carbonates of sodium and potassium which act by producing sparingly soluble compounds of the metal to be protected. For example, a precipitate of $Fe(OH)_3$ can be produced over steel surfaces by adding alkali (see Pourbaix diagram).

Vapour phase inhibitors are organic fluids such as sodium benzoate, cyclohexylamine carbonate which evaporate slowly and blanket the parts to be protected in vapour. It is usually necessary to wrap the components in some way and impregnate the wrappings in order to confine the protective vapour.

Cathodic protection

Since it is always an anode which is corroded, one way of preventing corrosion would be to artificially make the metal to be protected into a cathode. The Pourbaix diagrams indicate ways of doing this. Such protection can be afforded by placing the metal to be protected into electrical contact with a material having a more negative value of E^0. Thus, underground steel pipelines can be protected by using the sacrificial corrosion of zinc or magnesium blocks buried near the pipeline and connected to it. An alternative method is to feed a negative d. c. voltage on to the metal to be protected, the positive supply being connected to inert anodes buried near the pipeline.

BIBLIOGRAPHY

ANDREWS, E. H., *Fracture in Polymers*, Oliver and Boyd (1968)
BIGGS, W. O., *Brittle Fracture of Steel*, Macdonald and Evans (1960)
CHAMPION, F. A., *Corrosion Testing Procedures*, Chapman and Hall (1964)
EVANS, U. R., *An Introduction to Metallic Corrosion*, Arnold, London
EVANS, U. R., *Corrosion and Oxidation of Metals*, Arnold, London
FINNIE, I., and HELLER, W. R., *Creep of Engineering Materials*, McGraw-Hill (1959)
FORREST, P. G., *Fatigue of Metals*, Pergamon (1962)
Fracture of Engineering Materials, A.S.M. (1964)
Fracture of Metals, Institution of Metallurgists refresher course (1949)
GEMMILL, M. G., *Technology and Properties of Ferrous Alloys for High-temperature Use*, Newnes, London (1966)
KENNEDY, A. J., *Processes of Creep and Fatigue in Metals*, Oliver and Boyd (1962)
LOGAN, H. L., *Stress Corrosion of Metals*, Wiley (1966)
PARKER, E. R., *Brittle Behaviour of Engineering Structures*, Wiley, New York
POPE, J. A., *Metal Fatigue*, Chapman and Hall (1959)
SCULLY, J. C., *Fundamentals of Corrosion*, Pergamon (1966)
STEWART, W., and TULLOCK, D. S., *Principles of Corrosion and Protection* Macmillan (1968)
Toughness and Brittleness in Metals, Institution of Metallurgists refresher course, Iliffe, London (1960)
WULPI, D. J., *How Components Fail*, A.S.M. (1966)

CONVERSION TO SI UNITS

	SI Unit and Conversion Factor	
Ångström, Å	1×10^{-10} m	
ft	0·304 8 m	
in	0·025 4 m	
in^2	$6·451\ 6 \times 10^{-4}$ m^2	
in^3	$1·638\ 71 \times 10^{-5}$ m^3	
lb	0·453 6 kg	
ton	101 6·05 kg	
dyn	1×10^{-5}	
pdl	$1·382\ 6 \times 10^{-1}$	newton, N—Force
lbf	4·448 2	
ft lbf	1·355 8	
erg	1×10^{-7}	
cal	4·186 8	joule, J—Energy
Btu	105 5·06	
eV	$1·602 \times 10^{-19}$	
hp	$7·46 \times 10^2$	
erg/sec	1×10^{-7}	
kcal/sec	$4·186\ 1 \times 10^3$	watt, W—Power
ft lbf/sec	1·355 8	
Btu/h	$2·929\ 4 \times 10^{-1}$	
dyn/cm^2	1×10^{-1}	
atm	$1·013\ 3 \times 10^5$	
tonf/in^2	$15·44 \times 10^6$	newton/metre^2, N/m^2
lbf/in^2	$6·894\ 7 \times 10^3$	—Pressure or stress
kgf/mm^2	$9·807 \times 10^6$	
mmHg	$1·333\ 2 \times 10^2$	

$$10^5 \text{ N/m}^2 = 1 \text{ bar} = 0·1 \text{ N/mm}^2$$

Miscellaneous conversions include:

1·0 g/cm³	$= 1 \times 10^3$ kg/m³
1·0 lb/in³	$= 2·768 \times 10^4$ kg/m³
1·0 lb/ft³	$= 16·018\ 5$ kg/m³
1·0 kcal/g mol	$= 4·186\ 8 \times 10^6$ J/kg mol
1·0 lbf/in²	$= 68·95 \times 10^{-3}$ bar
Thermal conductivity, cal/cm²/s/degC/cm	$= 418·7$ W/m degC
Specific heat, cal/g/degC	$= 4·187 \times 10^3$ J/kg/degC

Constants:

Gas constant, R	$= 8·314 \times 10^3$ J/kg mol K
Avogadro's number, N	$= 6·023 \times 10^{26}$ kg mol
Boltzmann's constant, k	$= 8·617 \times 10^{-5}$ eV/°K $= 1·38 \times 10^{-23}$ J/°K

INDEX